# One Mo(

An epic journey fu

Fir

**ISBN:** 9798618626040

Requests to publish work from this book should be sent to:
david@one-moore.co.uk

Cover Design: Stuart Clitherow
info@stuartsgraphics.co.uk

Special Thanks To : Sonia Cameron, Sera Relton
& Christian Gordon

Printed by Amazon KDP Print

Connect with David & see more pictures
of his adventures at his website

-

**www.one-moore.co.uk**

**david@one-moore.co.uk**

www.instagram.com/djmoore07

# Table of Contents

2

# To Olivia, Jessica and Orla

Having you all as a geriatric dad meant that you missed the vast majority of my years within JLR. I realised that I had done so much that I wanted to share with you and I just had to get it down in print.

I dedicate this book to you for giving me a reason to write it. I hope you enjoy reading it as much as I have had living it and writing it.

# Acknowledgements

I have shared my time at JLR with so many great people that have made the last thirty years an absolute blast and I thank you all for your support. Without you all I could not have had so many experiences to write about! If you are not mentioned in the book you still all know who you are and I hope you still enjoy reading about our fun times.

I would firstly like to thank my mother and father who unfortunately are not with us anymore. My Mum always kept me on the straight and narrow and my Dad played a massive part convincing me that I was not quite good enough to be a professional footballer but I should take up an offer to be an engineering apprentice. Alex Ferguson will never know what he missed!

Massive thanks to David Steele for making sure I was in the right place at the right time and getting me an interview for a job in Vehicle Proving. I could write a book on our exploits together but we would end up in a right pickle. Those early years were very special and totally bonkers!

Guiding me along my way were a number of JLR legends including Mark Fletcher, David Hopkin, Vyrn Evans, Peter Fawke, Terry Igoe, Andy Whyman, Mick Mohan, Phil Hodgkinson, Kev Riches, Wayne Burgess, Joe Buck, Gary Hone, Graham Wilkins, Martyn Hollingsworth, Tony O'Keeffe, Peter Matkin, Tim Matthews and Pete Page. Without your help and support there would be no book!

A special thanks to Ken McConomy. A fellow Manchester United fan who beat me to the job of PR Manager but then ensured that I was involved in the vast majority of drives and launch events around the world that make up a lot of chapters in the book.

Also special thanks to Sonia Cameron for reading every chapter and giving me the encouragement to carry on and finish it.

Finally, to Danielle my work colleague, best friend and then wife. All of your love, support and advice throughout all my escapades really has been invaluable. Whilst overseas doing high speed testing or being away on launch events your understanding and support has been amazing. When I have been working at home until the early hours of the morning you are always there with a cup of tea and a cuddle.

You are my rock and I love you loads x

# Foreword by Jay Leno

At my garage I meet a lot of automotive representatives, usually they are marketing people or PR flacks sent to help sell the latest hybrid technology or explain why their vehicle is the class leader in their field. Most are quite good at their jobs having memorized key phrases like ergonomically designed for maximum comfort or no other vehicle in this price range offers these features! For most of these men and women it's a good paying job, for the men and women of Jaguar it's more of a calling and I think that's because Jaguar is an inspirational brand.

Unlike other upscale brands like Rolls-Royce or Aston Martin, Jaguar has always been attainable. It brought twin cam engineering and exotic styling to the common man. No other car company captures the pride and the soul of the English people like Jaguar. Whenever their employees come to my garage they feel more like club members than employees and David Moore is the epitome of this type of attitude. I first met David back in 2007 when he brought his Jaguar XF to my garage, from the launch event in Phoenix, and we had a ton of fun driving in the California hills as well as subsequent events at the Goodwood Festival of Speed and the Mille Miglia. David's 30 years with Jaguar have taken him on an epic journey, and his tales from around the world are both entertaining and insightful.

One of the thrills of my life was meeting the grand old man of Jaguar, the legendary test driver Norman Dewis so it's nice to see men like David carrying on that tradition.

**Jay Leno**

# Introduction

Jaguar winning at Le Mans in 1988 with the XJR-9 and then the launch of the XJ220 in Birmingham in October 88 changed my life in so many ways. Both cars had become my pin up posters. The scenes of the XJR-9 as it went over the finish line in first position gave me goose bumps and then seeing the XJ220 at the motor show left me speechless.

Having left it late for engineering apprenticeships in 1980 I missed out on Jaguar but ended up in a small company making commercial and aerospace components. I managed to do quite well in the machine shop picking up a couple of awards and was promoted into the position of draughtsman and cost estimator. Whilst I really enjoyed the aerospace side of the business, I was always wondering what it might be like at Jaguar. This was somewhat fuelled by Dave Steele who became a mate at college and who was working his way up through the company. He would tell me all about the development work he was doing and the trips around the world testing which sounded so glamorous.

Although I could not get to Le Mans Dave did manage to get some tickets for the motor show where he was keen to show me the XJ220. The crowds around the car were about 20 deep and it was causing quite a stir but Steeley knew all the right people and before we knew it we were on the show stand with a glass of gin & tonic and unlimited access to the world's fastest sports car.....I was in heaven and I made it quite clear to Steeley that if there was ever a chance to be part of Jaguar I would take it!

In 1990 that chance came knocking at my door. Dave was working in the Vehicle Durability and Testing Department and they had some vacancies coming up and were looking for non-automotive engineers that may have a fresh eye view of the business. I applied for a test engineer job and was then interviewed at the Whitley Development Centre and with all the coaching I received from Steeley I was starting to dream of what might be.

They advised me that I would hear from them in a couple of weeks, so I went off on a holiday to Florida with my fingers crossed.

I returned from holiday and my Dad picked me up from the airport. As we were driving home, he passed me an envelope that was clearly stamped Jaguar Cars. My heart was in my mouth......I opened the letter and read that I was being offered my dream job and after only two hours back in blighty I was sitting in front of my boss with a letter of resignation trying to look sad about handing it over.

So, the next month as I walked into Whitley past the statue of Sir William Lyons in August 1990 I had no idea of how this iconic company was going to change and shape my future. I was now part of the Jaguar family ready to start a new career and a new and exciting chapter in my life. A job that would enable me to travel the world, build and test some great cars, meet some great people, VIP's, sports stars and more importantly the opportunity to meet my wife and have three gorgeous girls who are one of the reasons for me writing this book.
It's a privileged journey of automotive passion with some fantastic stories that have shaped my life forever.
The journey has been eventful, funny and at times emotional and I hope all the short stories will be an inspiration for others, especially my children for who I hope I have left a legacy.

And to my girls remember this,  like a relay race I have now completed my leg, I learnt all I could from the race, I ran as fast as I could and I'm now passing over the baton for you to play your part and although I may be getting older I'm always here if you need me.                                          '

# Part 1 – The Start of Everything

Jaguar Engineering, Whitley, Coventry, and the beginning of a journey.

In Chapters one and two I go through my arrival at Jaguar Cars and a summary of the areas of the business that I have been attached to over my 30 years. I will briefly touch on how I got to where I am today as a background for all the worldwide tales and events that I have been on.

In order to get to these epic events there is a vast amount of hard work that has to be completed by a lot of dedicated people. Project teams are created for each programme and we have Design, Planners, Timing, Engineering, Quality, Prototype Build, Production Manufacturing, Public Relations, Sales & Marketing and Vehicle Launch all contributing to the development and final release of our amazing vehicles.

Before we even get to project teams our Research & Development areas have already logged up thousands of hours to ensure we are going in the right direction and our Competitor Vehicles department have benchmarked the relevant competitor vehicles that will help us achieve our targets.

It's only after a lot of hard work that we are finally able to launch and promote our brand, which then gives me the opportunity to do all of the things in this book.

I appreciate all the blood, sweat, heartache and sometimes tears that goes into what we do, and I thank you all for part that you have played. Enjoy the best bits!

# 1. A New Home

So here it begins, August 1990 and it's my first day at Jaguar Whitley and I'm full of all the normal emotions of any new venture. Apprehensive, nervous, excited and thrilled to be part of the Jaguar family.

Dave Steele collected me from the main reception and we headed down to my new home of Vehicle Proving. The department were responsible for the test and development of all the new prototypes around the world to ensure that the vehicles were robust enough in all conditions and then signed off to the appropriate standards. It was an area that engineers really wanted to be part of, so I was feeling very lucky to be here.

Vehicle Proving was split into several departments. There was Durability Testing, Specific Vehicle Testing, Competitor Vehicle Benchmarking and Resource & Data Management with the latter being my first assignment under my new boss John Trout. John was an interesting chap and a typical engineer. He was passionate about fixing everything in his path and as I would find out later in Timmins if something was broken or worn out, he would buy nothing new until he had exhausted all the possibilities to fix it first.

Having established where I was sitting and meeting the team I was then given the guided tour of the site with Steeley. Compared to my previous places of work, Whitley was an eye opener that blew me away. I had never seen so many computers and drawing boards in one place. I was then taken through Body Development, which must have been the length of a football pitch, Electrical Development, Powertrain Engineering, where all the ramps were occupied with prototype test vehicles for the yet to be released XJ Saloon and XJS Sports Cars and then over to the build workshops with lots of secret test cars being prepared for all sorts of development. It was far too much to take in, but I was assured that I would get used to it.

Interestingly the Whitley Engineering site used to be an airfield built in the 1920's, bought by Armstrong Whitworth and was the home of the Whitley bomber.

The site then closed in 1968 but was bought by The Rootes Group in 1969 for design and engineering of trucks and cars with Chrysler Europe before being sold to Peugeot in 1978.

In 1985 Peugeot moved its design and development to Paris and Jaguar bought the plant in 1987.

So, the entire facility still had that refurbished newness about it, and I loved it.

I had a new desk, a new computer, my best mate alongside me and the door was opening to a career and a life I could not have dreamed of or wished for that would subsequently make this book possible!

# 2. A Man with a Plan

Having established myself in Resource & Data Management and spent some considerable time looking at our operations I decide that I definitely need to expand my career at Jaguar. Whilst I'm busy looking at testing workloads, resource plans, timing plans and doing a lot of analysis of worldwide testing issues I'm yearning to go further and get involved a lot more on the testing side. I discuss this with Steeley and some of the elders in the department such as Bob Beebe and John Wood who were legends that will get a couple of mentions from overseas testing. Between them they decide that I need a plan of action to extract me from my desk and all the great work I'm now doing for John. Compared to lots of other jobs and my previous employment the role I had was amazing, interesting and rewarding but I would sit there and listen to all the conversations about testing cars and the stories attached to them and dream about doing the same things.

Without further ado I drafted up an action plan and requested that I should go over to do a stint at MIRA (The Motor Industry & Research Association) near Nuneaton where Jaguar had an onsite facility for durability testing. The elders believed that doing this would help me on my journey with vehicle testing and open more doors within the business....... and they were not wrong!

A month or two later I'm transferred to MIRA and I'm busting a gut to get stuck into everything they have to offer. It was also handy that I lived in Nuneaton, so the daily commute was a ten-minute drive with no traffic, unlike the drive to Whitley with all the masses.

A bit like Whitley MIRA started its life as an airfield better known as RAF Lindley. It opened its doors in 1943 as a training unit supporting other local bases such as RAF Bitteswell and RAF Bramcote and became the home of No 105 Operational Training Unit flying Wellington Bombers and Douglas Dakotas. My mother informed me that she remembered Sir Winston

13

Churchill driving into the airbase in a tank when she was a young girl and when she grew up she also had a job at Lindley Lodge which would have been part of the base which then became part of the coal board where she eventually met my father. In 1946 just after the war the site was sold off and taken on as a test and research facility which utilised some of the runways as part of its test track and even today still has the old control tower.

As with most newbies I'm assigned a mentor who in my case was a chap by the name of Mark Fletcher. Mark was ex-Army and from day one I liked the cut of his jib. He told it like it was and ran a tight ship in the durability office, but he also knew how to have some fun and was an excellent team player. We got on like a house on fire and it wasn't long before we became really close mates and he still is today even though he has now retired. When he told me he was retiring, and the Design Team invited me to attend his leaving presentation I knew we had to make the day special. His boss had already asked about the possibility of having a Jaguar Car to collect him from home on his last day and it was a no brainer for me to sort it out and be his chauffeur for the day. In my current role within JLR Classics I managed to secure a 1950's MK9 Jaguar from the collection and picked him, his wife Viv and his daughter Katie up from home and drove him into Whitley for one last time. It was emotional to say the least because Mark had supported me so much over the years and was one of my best friends for thirty years. Our trip to Timmins in Canada was the best and it gets its own chapter later in the book, so Mark has not gone quite yet. Thanks for everything mate you are one in a million!!

The 'Back Office' as it was referred to was a place for the engineers to reside and control test operations that would become my new home. I shared the space with some great engineers too. Nick 'Mad' Dog Gilkes was a relatively young buck who had come through the Jaguar Apprenteship scheme and was now running a fleet of test cars on both road & track durability. Steve Botterill was working in Specific Vehicle Testing looking after brake & performance testing, Ralph Shores was looking after Road Load Data from our structural test programme and then we had many visiting engineers on a daily basis doing all sorts of track-based testing.

I learnt so much from these guys and it wasn't long before I had my own test fleet of Jaguar Saloons running around the UK completing real world road durability. As part of the test I also got to drive the routes and evaluate the products but first I had to get my Jaguar Driving Permit.

To enable me to drive Jaguar Vehicles on the road I had to complete a Class 2 assessment by one of my seniors and in my case, I got Bob Beebe. As I mentioned earlier, Bob was one of the Vehicle Proving Legends who looked, spoke and conducted himself just like Sean Connery. For my assessment we took out a LHD XJS just to make things a little more difficult. We strapped ourselves in and set off from the workshop, through security and up to the MIRA Exit. As I lined up to turn left I asked "How's it looking Bob" ….."All clear mate" and as I went to pull out he said "After this one!" I hit the brakes so hard and looked at Bob who was smirking like a Cheshire Cat. My heart was pounding with the adrenaline rush and of course, there was nothing coming. Bob just fancied testing my reactions a bit and ticked off one of the assessment boxes.

Having passed the Class 2 assessment, and subsequent medical, I was all signed off and ready for action. In my small fleet of test cars I had a gorgeous XJS so Mark suggested that I should take that out on one of the better routes. We had all sorts of routes with different road surfaces, traffic, stop starts, hill climbs etc. but the route I selected was through scenic Wales.

There was a huge sense of pride with every vehicle I drove, and you just had to smile! It was a fantastic job. There I was in a two-seater sports car driving around Wales and being paid…..dreams were coming true!!

Having established myself in the role it wasn't long until I branched out onto the track, which was something I really wanted to do. We used to run a test called GAMA or General Accelerated Mileage Accumulation. This really put the engine and gearbox through its paces and allowed the powertrain engineers to monitor oil consumption. The test was conducted on the banked track at MIRA and required the cars to go flat out on the straights. Although I was running the test plan Jaguar had a pool of durability drivers that were cycled around the tests and a few of them were keen to do GAMA. Tony Cole, Dave Stew, Kev Brown and Ivor Cook were all on my list and what a great bunch they were. Tony "The Italian Stallion" looked every inch an Italian and learnt to speak the lingo from many test trips to Nardo in Italy. He also had a Harley Davidson and always looked the part. We both shared a passion for motorbikes and anything else that was fast. Dave "The Bookie" loved horse racing and as my dad worked for Ladbroke's we got on really well. Dave was a steady chap though and really focused on the test work. Brownie was the lovable rogue and always up to mischief. There was a very serious incident in my time at MIRA regarding a missing cake. The cake was made by the hotel we used in Italy and was sent for the attention of one of the other drives but when it was officially opened there was nothing left but crumbs. It sparked off an internal investigation, but they never found the culprit or culprits, but Brownie does like a bit of fruit cake!! Then there was Ivor the Driver. Ivor loved high speed driving and would volunteer for anything on the banked circuit. He was a bit older than the others and had bucket loads of experience. During one of my passenger rides on the banked circuit we were flying down the straight at about 130mph and he set the car up to hit the banked curve. Just before the car climbed the bank he knocked a pen off the centre console and asked me to pick it up. Now for those who have never driven at high speed on a banked track there is an element that you learn about quite quickly. It's called quite simply G Force. With aircraft, in a banked turn of 60 degrees lift equals double the weight or $L=2W$ so the pilot would experience 2g and a doubled weight. The same applies in a car hitting the banked curve. As you fly around from about 100 to 130mph you can feel the pressure of your body in the seat. So being a bit naïve I slacken my belt a little and bend forwards to pick up the pen just as we hit the banked track. Ivor accelerates and keeps the car nice and close to the armco, but I'm stuck. I have the pen but I can't get up. All I can hear is "Are you alright down there lad.....do you need a hand". He slows the car down and we both burst into laughter. I used that very same trick on several graduates over the coming years.

Anyway, this taster on the track meant that I had to get my next track permit a 3PG (Class 3 Proving Ground). The Class 3 was based around advanced driving techniques along with Proving Ground driving assessments but being in Vehicle Proving meant I had some great coaching and passed it quite quickly.

I have always had a love of speed that started when I was young. My father was never a speed demon and was always very cautious but fortunately my Uncle Ray was the total opposite. Ray was married to my dad's sister Jose. He had a garage business and he raced motorbikes at the weekend. When I was about five or six, he built me my first Go Kart that we loved pushing down the hill trying to go as fast as possible. It was a typical wooden carriage with big pushchair wheels and painted bright red. The steering was a piece of old rope anchored at each end of the steering bar that your feet pushed to turn. I think this must have been the start of everything so I must thank Uncle Ray!!

When I got to about seven, he built me a small motorbike that we used to ride at the local quarry. It came with a rope attached to the back and a handle for him to hold onto so that we could not ride off into the distance. Before long the rope was off and I was addicted.

My father moaned quite a lot about fast cars and fast bikes, but he never stopped me doing it as long as Uncle Ray was involved. As I grew older, I then started watching Ray at race meetings on his TZ250 Yamaha and it became a bit of a family gathering that I enjoyed so much.

The noise, the smell and the occasion really gets to you. He used to let me freewheel the bike off the track and back to the van and I really thought I was special. I guess it was only a matter of time before I then wanted to race but I think both my Dad and Ray decided that I needed to get some better motorbike handling before being allowed out on the track. I was introduced to trials riding next where you try and go around a course without putting your foot down at really slow speed. The art is to master bike control and balance. All the good riders were using lightweight Montessa bikes so when they said I could have one I thought that's what I would get. As it happens, I ended up with a heavy old beast called a Bultaco Sherpa T which had previously been owned by a chap called Sammy Miller who was a famous trials rider in his day. On moaning about its weight I was told that if I could handle that I could handle anything. They were good words of wisdom and I played around on it for a couple of years gaining some good bike control.

During that period I also started looking at road bikes and after I left school I was keen to get some wheels instead of getting the Midland Red Bus to work. Rather than buying an old second bike I just had to have the latest and fastest thing on the market, so deals were done with Team Ham Yam and I purchased a Yamaha 250LC. It was the bike that changed motorcycling overnight and became the weapon of choice for many road racers in its day. It was also the first 250cc bike to go over 100mph and was water cooled. It felt so good to ride and I was in my element. Obviously over the years I had many bikes moving though the bands to 550cc, 750cc then 1100cc getting more powerful and more exciting. Uncle Ray at this point had taken up an additional hobby that I also became interested in....Flying!

He managed to get his pilot's licence and started flying a Cessna out of Bagington Airport in Coventry so why wouldn't I want to start flying? I did a few flights with him and guess what.....I got addicted to that as well. It was kind of expected really because when I left school I was desperate to be an RAF Pilot. I wanted to fly Tornadoes as they were the latest and greatest in the RAF replacing the Lightning as our front-line fighter.
Unfortunately, my eyesight was not good enough and the options of being a navigator or flying Hercules aircraft were not on my wish list! It was not meant to be.... a missed opportunity........or maybe not, as I will write about later.

I did start my pilots licence but the costs were mounting and I never finished it, but one day I will go back to it. The quest for speed though finally got me onto the track some years later whilst I was at MIRA. Good old Uncle Ray agreed to let me ride his latest TZ250 and I got myself a race licence and made my way to Mallory for a few afternoons on the track practicing. I managed to do a bit of racing but as I had lots of other interests and pursuits I had to give it up as it was an expensive hobby.

Still, I was now the holder of my Jaguar Class 3 PG Permit that gave me unlimited access to our own private test track 5 days a week. My aim now was to gain as much track time as I could and learn as much as I could from the test drivers and proving engineers as I wanted to be up there with the best. I was quite lucky really because Mark Fletcher was a brilliant mentor and also a great driver who guided me through my high-speed testing.

I was soon doing my own mileage accumulation tests and high-speed specific vehicle tests. There was nothing better than being on the No 1 Banked Track flat out in a Jag. As part of the process when you test cars you go out and learn about it slowly. You take it steady at first and always build up your speed and trust. You also do regular ramp checks, tyre checks and report out any issues that you find and then when you are happy you start getting quicker. The experience is quite amazing and as near to American NASCAR or Indy Racing that you can get. When I was ready, I always pulled out onto the circuit and did a couple of warm up laps before moving over to the outer lane. The circuit we used had three straights and three banked curves of varying lengths. The XJ and XJS cars that I was testing were capable of 155mph but the restrictions on the banked curves reduced that to between 100 and 130mph max but at that speed it was exhilarating. After flying down the main straight you have to let the car ride up into the bank then ever so slightly turn the wrong way into the bank so that the car can ride up to the top. Once by the armco you relax your grip on the wheel a bit and focus on steering around the curve using your throttle. A slight increase to climb up and back off a bit to come down.

I never got tired of doing this and I even got to the stage where I was doing demo laps and VIP tours which were so much fun because the passengers' reactions were priceless. One such VIP was Martin Johnson, who at the time was then England Rugby Captain. We were actually doing a site tour for the Leicester Tigers, but a local radio station wanted us to do a few laps and interview Martin on the way around. As the interview started, I launched the car into the banked curve and they managed to catch a full-blown scream from the big fella. He was a gentle giant to be honest and seeing him lift the World Cup was just a bit special.

I was really enjoying the high-speed testing and on one particular day the competitor vehicle guys turned up in a new Porsche 911 which I really wanted to drive, unfortunately this type of car required the next level of permit at Class 4. I was straight on the phone to Dave Steele and within 24hrs I was booked in to do my 2-day assessment, which in those days was called your HPC or High-Performance Club. The assessment was conducted by a gentleman by the name of John Lyons who was an ex Police Driver Training/Instructor. The assessment would be a series of road-based drives, a track drive at Millbrook Proving Ground and then a racetrack assessment at Cadwell Park.

I was well up for it, but a lot of my fellow engineers were telling me that he was ruthless and just don't argue with him at all. Dave Steele had two bites at passing this test and on his first go John sat him down at the end and said, "You look like a cowboy; you drive like a cowboy and my god you are a cowboy…Failed".   Apparently they just wound each other up for two days and Dave lost interest, but it worked out ok on his second go.
I decided that I would be respectful, listen and do what he told me to do and smile a lot. We set off from MIRA heading south to Millbrook and spent most of the journey getting to know each other. When we arrived we did a few vehicle checks and headed out onto various tracks for some 'on the limit' handling tests.   With all the MIRA track work I had been involved with

over the previous few months I found this part of the assessment quite relaxing and enjoyable and I didn't want it to end!! When it did finish we had a quick review and John advised me that we would be heading north to Cadwell Park for an overnight stay, but on route to the hotel I would be assessed for my road based driving and also complete a commentary drive as well.

With a commentary drive you have to describe in detail everything you are doing as well as commenting on what you can see, observations in road conditions and anything that may hinder your ability to make progress unobtrusively. It was the most intense driving I have ever had to do, and John was pushing my buttons all the time to see how I would react.

After an epic fast drive, we arrived at the hotel and I was shattered both mentally and physically. I completed a post drive check of the car and then John suggested we meet in the bar for a debrief in about ten minutes. After all the hype from my teammates I was more than a little nervous as I walked into the bar, but I needn't have worried as he was quite keen to get a drink and some food and the review of the day was quite good.

The next day we set off for Cadwell Park and I was pleased we were going there as I had spent a bit of time watching bike racing here and even completed a couple of track days, so I knew the circuit well. When questioned about the circuit by John I told him that I had good knowledge to which he replied "Excellent, then you won't mind doing the course anti clockwise then". Well…it was a challenge I have to say, but I did as I was instructed to do, and I had a blast. We left the circuit and headed back to MIRA with some more rapid road driving with full commentary.

To give you an idea of what that commentary is like the following is a very brief couple of minutes on a B Road in the countryside:

*Ok, checking road and mirrors, all clear so accelerating and selecting appropriate gear. Looking ahead to plan fastest and smoothest route. Checking forward tree lines and telegraph poles indicating road is going to the left with slight incline. No oncoming traffic so positioning the car over the centre line for better line into left hand bend. Checking for farm gates and side roads but no visible indications. Road straight ahead so accelerating to make progress. Road surface is good and dry so no current concerns on tyre grip but I am cautious of possible mud or farmyard debris at any point.*

*Oncoming low rising kerb indicates possible dwellings and pedestrians so braking to reduce speed. Road signs indicate small village coming up so speed brought right down. Light oncoming traffic and lack of pedestrians. Watching side roads for vehicles pulling out. Making sure car is positioned suitably and in correct gear for possible acceleration.*

*Dustbins are visible for possible collection so anticipate waste disposal truck on either side of the road.*

*Continue through village but noted two people at bus stop on right hand side. Possible oncoming bus on exit of village so proceed to accelerate with caution.*

*Oncoming bus noted. Clear of bus, road conditions clear so accelerating to make further progress.*

This is how it went for a couple of hours and by the end of it you are completely drained.

As you were doing the commentary he would also ask you questions like "What did that caution sign just say?" or "Can you tell me the name of the village on the previous right turn?"

It was intense but rewarding.

As we approached MIRA he suggested we stop in at the Café near Hinckley for a cuppa and a review of my two days. The word Cowboy was in my head from Steeley's first review and I wondered how he viewed my driving ability. He mentioned areas to improve in and also gave me some tips, but his final summary was it was a good display of swift motoring with improving track skills and he could recommend me as a member of the High-Performance Club. I was thrilled to bits and that cuppa tasted mighty fine so cheers John!

So I was now armed with all my permits and some excellent durability testing under my belt and I felt like a good Vehicle

Proving Engineer. I was now ready to tackle the overseas assignments that you can read about later in 'Tales from around the Globe'.

I was now a man with a plan and I had my sights set on another area of the business that I hoped would open more doors within Jaguar.

Whilst I had been doing all my MIRA activities Steeley had moved into Competitor Vehicles and was looking at moving higher up the ladder into project management via the programme office and as such he felt that it would be a great move for me to come in and work with him for a while and the idea was to hand it over to me at some point. His advice was clear and simple. This role will give you more visibility in the company than anywhere else and he wasn't wrong.

My new boss was a chap called Dave Hopkin who had taken over durability testing and competitor vehicles. Dave was ex Road Load Data and without a shadow of a doubt one of the nicest blokes in Jaguar who was just one of the boys at heart. In my years working for him he sent me all over the world doing some amazing and slightly crazy things and we had an absolute hoot. We worked hard, we always delivered our objectives but every day we laughed at various situations that we ended up in.

The 'German Barrier' was one such situation that we used as an example to others who came in to work for us about how even with all the best laid plans things can and often do go wrong. So, the German Barrier episode happened, obviously in Germany, when we were holding a company ride and drive with all the execs. Hoppy decided that it would be a great idea to get a coach and collect all the directors from the airport and drive them back to the hotel in one go. The coach was sourced and arrived in good time and off it went to the airport with Hoppy on board. On arriving the police asked for it to move along and eventually Hoppy decided that it could go into the short stay parking. The plane arrived and as the team were coming through control Hoppy instructed the coach to pull out of the car park and circle around to the pickup area. Unfortunately the barrier had broken and an engineer was sent for quickly who promptly arrived and set to work to repair the barrier and let the coach out. The team not wishing to hang around decided to walk to the car park and at least get on the coach and be ready to leave straight away. It was a schoolboy error but Hoppy had lost his rag with the engineer and was giving him an ear bending about the time taken and he had a coach full of execs to get to the hotel ASAP. Well....the engineer had the last laugh as he downed tools and just walked off leaving the coach and its passengers stuck for a little while. For years afterwards on any of our events we always remembered 'The German Barrier".

Having been tucked away at MIRA for a few years the Competitor Vehicle role brought me right back into the action at Whitley and as Steeley suggested opened all sorts of doors that the vast majority of engineers would never get a chance to do. I was lucky to be in the right place at the right time with the right mates looking out for me. I guess the best thing about the department was that it always had the best cars being brought into the company for benchmarking and when the chiefs noticed them on site my phone would be ringing all day!

Suddenly without too much effort, most of the Whitley Execs, Chiefs and Senior Managers know who you are and that helps on many levels. What would happen is that I would get loan cars in from Ferrari, Porsche, BMW, Mercedes & Lotus so that engineers could assess them against our own models but the execs would want to take them on overnight assessments. They had no time during the day so they could drive them home and bring them back the next day. It was a very rewarding part of the job too because the execs would often let me use their Jaguars from time to time.

The new role also opened the doors to programme Ride and Drives in various countries, doing different track testing and being asked to support events with the PR team. I threw myself 100% into all these opportunities with a passion and fortunately it didn't go unnoticed resulting in some once in a lifetime events that I will come to later on.

One of these events was a ride and drive in Atlanta with our Ford Colleagues as we assessed the Lincoln & S-Type prototypes against Lexus, BMW and Mercedes. I was asked to be part of the team to run the drive assessments and at the end of the event another door opened into the Vehicle Office that I could not miss. Competitor Vehicles had been good for me but now I had a lot of key skills that could be used by a project team and I needed to learn more about project management.

On returning to the UK I was soon drafted into the S-Type Vehicle Office working for Vyrn Evans. Vyrn was another one of the Jaguar legends and a top bloke. He was quite a small chap from the Welsh Valleys but had a big heart and wealth of vehicle knowledge from Land Rover and Jaguar. Although this was a great challenge for me there was some sadness as I was going to be looking after the project on my own and stand in for Peter Fawke who was taken ill with stomach cancer. I had met Pete in prototype build and also on the Atlanta drive, so we knew each other really well and it was a shock to the system when I found out about his cancer.

Fortunately, his treatments went well, and he recovered and returned to work and became an absolute inspiration to me. He came back as my manager and I was thrilled to work for him.

As a result of my testing, driving and PR events work, Peter always allowed me to be utilised on all sorts of events and he pushed me forward for lots of extra activities that he could have done if he had wanted to.

As I write this Pete has also decided to leave the company after many years and taken early retirement and although our work paths have drifted apart over the last two years I will miss him being just at the end of a phone for a quick chat, a gossip or some advice. Without his support there are some big pieces of this book I would not have been able to write so cheers mate and all the best (even if you are a blue nose)!!

S-Type Vehicle Office also threw me a curve ball that would change my life in many ways. Pete Fawke had taken on a new graduate and he was keen for me to pass on my knowledge and show them what we do and get involved in a few tests. What I didn't realise is that the young blonde engineer that I had been admiring for a week in the office was the very same graduate he wanted me to take on. I was introduced to Danielle and I have to say I was like a puppy dog at first, but I pulled myself together, wore my best suits and polished the Breitling watch to make sure I impressed her. It wasn't until some years later that she admitted that the first thing she noticed about me was my watch and she was a big fan of the brand (Lucky me eh).

Well as it happened, I had a specific test that needed to be done to calculate the real-world fuel economy of the new S-Type and we had devised a suitable road route for this and had two cars ready to go. This would be a perfect chance to get out and away from the office to chat with Danni.

With full tanks in the cars we set off and headed north to Snake Pass and Derwent Water where I knew there was a great place for lunch. We stopped at a village to get some food and I should have offered to get hers, but I missed the chance that she never forgot and still reminds me of it to this day. Well the walk at the reservoir and lunch must have gone well because over the next couple of months we became a lot closer and within a couple of years we got married and then had our three gorgeous girls, Olivia, Jessica and Orla who were the inspiration for me to write this book.
Danni still works for JLR and followed me into Special Vehicle Operations where she is having a very successful career of her own and as I type this chapter she is currently on maternity leave with Orla.

Working in the Vehicle Office had now put me in a position to project manage my own programme which just happened to be the new S-Type. My roles included planning the builds of the cars, their subsequent testing and trying to make sure all the other areas of the business got access to the cars they needed. Over many years all the departments ran their own vehicle testing and ended up going all over the world doing the same things as the other areas hence the role of Vehicle Office was born. I used to tell people that the Vehicle Office role acts like the hub of a wheel. It holds together all the spokes which in turn are connected to the outer rim that allows progress to be made. The spokes were the individual areas such as Powertrain, Chassis or Electrical and the outer rim was the business.

The new role through the launch of the S-Type opened further doors into the world of PR and Press, and with some great managers behind me I was allowed to get involved in all sorts of extra activities like racing our prototype S-Type R around Silverstone against our Formula 1 car, doing a magazine article about the art of camouflage and also being sent to Australia to conduct a 10,000 mile test in ten days. I was loving every minute of it and I think most people admired my enthusiasm and passion for the brand.

One person who certainly did was Andy Whyman who had worked his way up through Powertrain Development and then been seconded to Detroit for a while on an overseas assignment. He had now taken over the role of Vehicle Engineering Manager for the new Jaguar XF Saloon.

The XF was going to be something really special. After the sad loss of Geoff Lawson, our Designer who did the S-Type, we were going to see something different from the new team under the leadership of Ian Callum and I was looking forward to being part of it.
Right from the start you could tell that the team was also going to be something special under the Programme Leadership of Mick Mohan.  Mick had been a Jaguar lad working his way through Body Engineering and was well respected through the business as somebody who knew how to manage and get the best from his teams. In my working time I have come across bosses and leaders and Mick was by far the best leader!

The bosses only ever wanted the right answer to help them progress their career. Decisions were not made to benefit the product or the business, they were made to protect themselves and their CV. Leaders were a whole different breed. They would work with you, guide you, praise you and the team, and always made sure that a pragmatic answer was available for any questions or issues. There were two occasions that spring to mind when I think about Mick. The first was when we had to have answers to some difficult questions for the project to move forwards and Mick stood there and said "Right, this is what we are going to do". I asked him how he did that and how did he know how to make the right decision? He said "I don't know if it's the right decision or not, but the project needed a decision to move forwards. If it's right, then great but if it's wrong then we review it again and make the right decision". I have used the same strategy so many times and it works. The second was when I stood up in a review and managed to dig a hole in a presentation I was doing. I was almost heading for disaster when Mick stepped in to say that he would like to discuss this some more after the meeting and don't worry about it for now. I sat back down relieved that I had got out of a tricky situation. Afterwards I went to his office and was greeted as nicely as always and then he proceeded to give me a dressing down for my poor presentation but then he advised me how it could have been avoided and suggested ways to improve for the next time. We then continued our chat over a cup of tea and it was all forgotten about.

This is what made Mick special and was also the reason that the XF Team set a benchmark for how programmes should be managed.

With this sort of leadership my time on the XF was just the best. I was also now one of the more senior members of the team so I was also helping out looking after new graduates and some of the younger engineers. I have to give a mention to young Matt Eyes who joined me on XF and became just the best wingman you could ever wish for. Without a doubt he was a breath of fresh air and together we had a blast. There are so many stories I could tell at this point, but I just can't fit them in (honest). However, two brief stories are worth a mention just for the giggle factor.

We always liked the odd prank in the office to keep each other on our toes and when I came upon a substance called pure capsicum, I just knew that we had hit the jackpot.

Darren Marcham from timing managed to source a tin of this stuff which you add to a curry to make it hot. You only needed the tiniest amount, maybe the size of a pin head, to really pep up a curry. So, in the interest of engineering and data collection we decided to play Russian roulette with meat samosas. We would take a pin that had been stuck into the capsicum and then push it into one of the samosas. You then waited until they had been shared out and just watch as everybody took a bite. The unlucky one was then seen running to get a cup of chilled milk. I adapted this by smearing the pin around Matt's teacup as I was making him a brew. I put the cup down and carried on at my computer. As I took my second or third gulp of tea Matt had his first and within seconds he went off like a firework. I spat tea out like a soda siphon and all those in the office who knew what I had done were in pieces.

Matt took it really well but had that glint in his eye that meant there would be revenge. It came a few weeks later when I just happened to mention that there was a strange smell in the office or more to the point near me. I was convinced that it was a dead rat somewhere, but I checked all over the place and could not find one. The smell got worse day by day and was getting to the point where I was going to have to call in for some help. Matt was offering all sorts of advice along with others in the area and no matter how much or what type of air freshener I bought, the smell overpowered them. At the point where it was getting just too much I got under the desk to check out under the floor tiles, but I noticed something taped under my desk with black gaffer tape. I removed the foreign object to find out it was a boiled egg that Matt had put there weeks ago that was now in a serious state of decay. The stench was so bad it made you feel sick but I acknowledged his efforts and we called a truce!!

The other short story turned into something quite epic really. Matt and I had a colleague in Vehicle Packaging called Martin Wheeler or Wheelie Bin to his mates. He had a new prize possession for the office in the shape of a lifesize cardboard cut-out of Monty Panasar, the rising star in the England cricket team. Now Martin was a massive fan and Monty was present at launch events, photo calls and the odd meeting just for a laugh so, one bright day Matt and I got scheming and decided that we should kidnap Monty and demand a ransom for him. We made up a ransom note and left it on his desk. Martin arrives back and instantly starts accusing us of hiding Monty and that we should grow up and give it back. Well this was like red rag to a bull and we decided to up the conditions a bit and sent him another note demanding a box of samosas for the office or Monty would get it. This fell on deaf ears as far as Martin was concerned and he just wasn't playing along. Our next effort did the trick though when we sent him a note from Monty saying he was having a blast and did not want to come home. We accompanied this note with a photo of Monty in bed with two girls who were friends of Matt. We were just plotting our next adventure for Monty when I read that the man himself was doing a book signing in Leicester and by a million to one chance, one of the chaps in the package team who worked for Martin actually knew Monty from college and had been keeping him updated on the "Where's Monty" Campaign. We just had to go so myself, Matt and another chap who got involved call Michael Rodley took the cardboard Monty to meet Monty in the bookstore. It was a surreal moment as we all stood there with the great man and he signed the cut out to Martin. It was eventually returned to its rightful owner and is now enjoying a peaceful life in the countryside near Leamington!

Well XF was a major success and as the Vehicle Project Engineer I made sure I was involved in everything taking on some additional roles that will take up four chapters in the section called 'Events Of Biblical Proportions' including a land speed record for Jaguar, a part in a TV show, helping out in a movie and meeting one of the most famous faces in America which is not bad for a lad from Nuneaton.

All good things do come to an end though and it's time for new challenges. With all my experience and the contacts I had made over the years I was offered a position to be part of a new group called ETO by a manager I knew from his Road Load Data days by the name of Kev Riches. ETO or 'Engineered to Order' was going to be a new department that would do just the Special Products. I had known Kev for a long time and done some work with him in durability testing so it was an easy decision which came at the right time as well because XF had run its course and I'm not sure where I may have gone.

Anyway, we get things going and before too long I have my first project called XKRS. This car was going to be our flagship sports car and my chance to do something a bit special. From my early days and my Le Mans passion I was a bit keen to do a sports car…who wouldn't? I had grand visions of taking Jaguar back to Le Mans in GT Racing and reproducing the magical times of 1951 to 1957 when Jaguar ruled the world with those historical wins at Le Mans with the C-Type and D-Type.

The move into ETO was a good choice as I was given the role to develop the Jaguar Sports Cars side of the business. I picked up two projects straight away with the XKRS and the XFRS from 2009 through to 2012, both of which attracted a lot of media attention and gave me more PR Events to do as well. With these main projects underway I still managed to do a lot of background work for our American Race Team Rocketsports Racing, who I had worked with on our XFR Bonneville Car and were now using the XKR in the GT Race Series. They were flying the flag for us in the USA and I was always ready to help them with anything they needed to keep the car going along with my Powertrain guru Dave Warner who had also helped me a lot with the Bonneville Project.

My continued support and enthusiasm were obviously appreciated because in June 2010 one of my dreams was about to be realised when I got the call to say that they would like me to be part of the team for the 2010 Le Mans Race. Having joined Jaguar on the back of the 1988 & 1990 Le Mans wins here I was being invited to participate as a team member. I didn't have much time to get ready and soon found myself with bags packed waiting for my chauffeur to drive me down. Who else would be better for a road trip like this other than my old mate Steeley. Dave was now in Marketing as a senior manager and was attending the event in a working capacity. We discussed our joint venture and decided that we just had to drive down and make it a road trip. We sourced an XK Convertible and managed to drive all the way there and back with the top down just like Thelma and Louise. The boys were back!!

When we arrived, I hooked up with Paul and John Gentilozzi and went straight to the pits. Unfortunately, the car was not running very well and had an intermittent electrical gremlin that we could not sort out. The only option was to just risk it and start the race the next day.

The atmosphere was amazing though and I couldn't believe that I was there. On race day I was bursting with pride when I helped push the car onto the grid for the start. It's an epic moment to be honest and I was so close to tears. I never pushed or intended to be in this moment at all.

It was achieved through a lot of hard work, being passionate about what I did and having a can-do attitude that was rewarded with this opportunity. Unfortunately, the race didn't last long for us when the gremlins attacked and we had to retire from the race. It was a crying shame for the whole team who had put in so much time and effort just to get over from the USA. We packed up all our kit with the intent to watch the rest of the race, but my heart just wasn't in it. I hooked back up with Steeley and we decided that an early run back to the UK in the morning was the best thing for the two Jag lads. We still enjoyed the drive back though and managed to have a few hours of gossip and reminiscing about our last twenty years!

Having returned from Le Mans I was still a bit down hearted but it was short lived. I had an email from the RSR team inviting me over to see them race in the Petit Le Mans at Road Atlanta in Georgia. I had read about this race and was thrilled to be invited.

When I got home, I suggested to Danni that we should fly out and spend the weekend there for the race. Her suggestion was, why not go out for longer and do a road trip. Hell yes! I sorted out the logistics with the PR Team and RSR advised that they would sort out our accommodation. All we had to do was get the flights booked and plan in a road trip. Our plan was to fly into Atlanta, watch the race and then head up to Washington to see

32

friends, then over to New York followed by a trip up to Boston for good measure.

We arrived, drove to the hotel and our first impression was OMG we can't afford this. We were taken to our room and we stood in silence for a moment just to take in the grandeur of a room bigger than our own house. "Don't worry babe" I said, "I have a credit card to pay" It really was a fantastic setting and for the first time in ages we could actually relax and enjoy some time together. Both of us had been really busy at work and we had also got married in March and for the last 18 months or so we had also been trying for a baby with no luck.

In the evening we met up with Paul and his wife and some guests of the team, but we were also joined by our chairman Mike O'Driscoll. The evening started with light discussion about Jaguar and I found that I could contribute to the conversation and enjoyed the banter, but then it progressed to a level where for the first time I was left unable to contribute to the conversation. They were all discussing which aeroplane to buy next, I mean this is the lad from Nuneaton who is thinking about a new Ford Focus and you guys are talking about the deals for a King Air aircraft at about $8M dollars. I was so out of my depth and considered telling them that I was thinking about getting the very latest Eurofighter or F35 in a deal with Airfix instead. We decided to retire from the conversation when we found out that Paul had an apartment in Trump Towers which might have been a tad swankier than our semi-detached in Nuneaton.

The race and the team were amazing and we had so much fun. We were also gobsmacked when we checked out to find that the bill had been paid. My credit card was also very happy! Thank You RSR!

Nine months later we were also blessed with a little girl called Olivia. Maybe we should have called her Atlanta?

ETO continued to expand with both Jaguar and Land Rover getting the full treatment and I was already working on the XFRS Wagon, but as always a door was opening for my biggest project yet to make the XKRS a better track car. The XKRS-GT was about to be born (see chapter 17). There were also a couple of little PR events around the corner with the most famous football manager in the world and another dabble with a famous TV show also featured later in the book.

The department was now growing at a fast rate of knots and had changed its name to SVO or Special Vehicle Operations. Kev Riches had moved over to our new Classics Division and I started a new challenge with Andy Smith as my new manager and Stuart Adlard as my Senior Manager.

I kicked off Project 7 as a limited edition two-seater and then moved onto F-Type SVR which would be Jaguars fastest production car capable of 200mph. Testing this car was something special because it looked perfect and the noise from the exhaust was fantastic. I had the pleasure of taking an F-Type on the Mille Miglia race in Italy and had a proper moment in a really long tunnel where I gave it the full nine yards of wide-open throttle. The noise was an audio extravaganza of engineering excellence that would not go unnoticed. As I came to rest at the toll booth two Pagani Zondas raced up next to me with the drivers screaming "Ballisimo" They obviously liked it then!

Having launched the SVR I then started Project 8 that was to turn a standard XE into a full-blown Nurburgring Track car.
We also had a change at the top and our new director was Paul Newsome who had been at Lotus and previously had a stint at Williams F1 working with Jaguar on the CX75. He came in with fresh eyes and it was clear that he wanted to move SVO in the right direction.

We had a few meetings to discuss projects and he asked us all for ideas to promote the department and lift the brand. Having dabbled with the Bonneville Project, completed a GT and kicked of Project 7 I could sniff the sweet smell of a petrol injected challenge. Paul was keen to do something to demonstrate the capability of the SVO team and I presented him with a presentation to take on one of the oldest and most daunting races in the USA…PIKES PEAK. A skunk project called FPP15 was born (see chapter 19) later in Events of Biblical Proportions.

During the latter end of 2016 I got a call from Kev Riches that would see me leaving SVO to join JLR Classics where they wanted me to create a bespoke XJ6 for the Iron Maiden drummer Nicko McBrain before turning my hand to create a reborn Range Rover Classic and then being asked to project manage the re-creation of a vehicle that would take me right back to where and why I started at Jaguar. If I'm lucky I may even do a second book for that one.

So…..this has been a brief review of my career path in JLR to date. I have made a lot of great friends along the way, met my gorgeous wife and I can't remember making any enemies, I have worked hard, travelled around the world testing, played a part in product launch events, driven hundreds of cars, appeared on TV and Film, made promotional videos and met so many famous people who all share our common love for cars. I created my own Jaguar GT, engineered a race car and set a record at Bonneville but I didn't do it on my own. I did it all with my best mates and we had a blast doing it. At times it was hard, but we had fun and we laughed a lot doing it. The following chapters have captured some of the most amazing times I have had at JLR and I hope you find them funny and interesting.

# Part 2 – Testing Tales from Around the Globe.

I have been fortunate to have been sent all around the world to do a lot of stuff.

My role in Vehicle Proving enabled me to conduct durability and specific testing in places like America, Australia, Canada, Germany, Mexico and Sweden.

My specific vehicle testing has also enabled me to drive on various test and race tracks around the world which I loved with a passion.

The next seven chapters will give a brief insight into the hottest, coldest, fastest and scariest places I have been testing and the fun we had doing it.

# 3. You're going to Timmins

It's my first overseas trip with Jaguar and the 8hr flight into Toronto was very relaxing in the company of Mark Fletcher and Ray Angliss from MIRA and Rob O'Toole from aircon development. We made our way through Toronto airport for our connection flight to Timmins. We finally found our booking desk tucked out of the way of all the mainstream desks and started to load up all our gear onto the conveyer belt. When the booking chap got to us and asked where we were going we said Timmins.......with eyes wide open and a chuckle in his voice he came out with "You're going to Timmins?" Maybe he knew more than we did but when we told him what we were doing for four weeks he sincerely wished us good luck.

One of my jobs to do on the trip was to deliver and test some engine management chips called EPROMS that were programmed in the UK for use on the V12 engines we had in the cars. Each chip had a different calibration setting and I had a rather nifty carrying box for about ten of these. I also had quite a bit of official paperwork to support the use of them that was to be presented to the customs teams if required. When I reached customs, I was asked to declare anything and I handed over the EPROM case and popped it open. The guy at the desk raised an eyebrow and questioned the contents. Now at this point I should have just handed over the paperwork, exchanged pleasantries and been on my way but oh no. In a somewhat naïve 'this is my first trip abroad' moment I uttered a few words that you really shouldn't to people who don't have a sense of humour. "Well, I could tell you but then I might have to shoot you". That line was perfect for James Bond and any other secret service agent but for me it was just stupid. What was I thinking? I guess I thought he might just see the funny side but alas no he didn't. We had to run through every bit of paper and go through in detail the ins and outs of my trip. The delay almost cost me my connection to Timmins but at least I learnt something from the experience....Don't joke with American Customs because they don't do funny!

So where was I heading? Timmins is a city in northern Ontario, Canada and it's situated on the Mattagami River. Its population is approx. 42,000 who support industries related to the mining of Gold, Copper, Zinc, Silver & Nickel as well as a thriving lumbering service in the summer months. It has a famous singing star that was born and raised there by the name of Shania Twain and its geographical location near the northern periphery of the hemi boreal humid continental climate (I had to look that one up by the way. It means it's halfway between the temperate and subarctic zones) means it can get warm in the summer and it will get very cold in the winter. The average temps in the winter are about -23°C and the record is -50°C which can feel much worse with wind chill. The reliable cold climates made it a perfect place to conduct CET or Cold Environment Testing for Jaguar from December through to the end of March.

Having enjoyed the space and luxury of British Airways Club Class on the 747 to Toronto I was now safely aboard a Dash Eight Turbo Prop with my other thirty nine passengers. Of these 8 were from Jaguar Cars and the other thirty two were mainly French speaking Eskimo Indians. My seat near the front gave me unlimited access and conversation with both the pilot and co-pilot through the open cockpit door. It was a little intimate in the cabin made worse by the artic clothing everybody was wearing except me. I had the optional denim jacket over a sweatshirt look which looked great before I left but now looked out of place on the arctic express.

I was advised before I left to purchase my cold weather gear in Timmins as it was better quality and cheaper than in the UK but my teammates had all been before and retained their arctic jackets for future trips. By the time we were airborne it was dark and had started snowing.

I cast an eye out of the window and could see the warm lights of Toronto disappear into the darkness to be replaced by just the flashing strobe lights of the wing tips. It made you feel a little bit uneasy as it was right out of your comfort zone as far as flying goes and the whole situation was made worse when suddenly there were a number of loud thuds against the aircraft. The pilot came onto the intercom to advise us that there is nothing to worry about as its only ice forming on the propellers and getting thrown off. Well that just made me feel a whole lot better....Not!

The flight lasted about one and half hours and it was still snowing as we circled Timmins airport and prepared to land. Now it might have just been my imagination but as I looked around the cabin, I'm sure that it was only the Jag lads who had their eyes open and fingers uncrossed. I guess the locals had done this trip before because the plane bobbed up and down in the snowstorm and the engine rpm was going all over the place as the pilot worked hard to keep it all level. The lights of the runway loomed large out of the darkness and we skidded down the snowy runway and came to a stop just short of the main building. There was a ripple of applause for the pilot from all the passengers but for me he deserved so much more!

The doors opened and we made our way down the steps for the short walk to the terminal and OMG how F in cold is this? The arctic clad souls made their way to the terminal without a shiver, but Mr Denim Man was seconds away from hypothermia. I have never felt cold like it.

We quickly made our way into the arrivals hall and waited for the truck to bring the luggage over when I heard a familiar sound bellowing across the hall. "Oi Moorey, welcome to Timmins". It was my boss and Team Leader for this trip Gary Hone. He took one look at the denim jacket, laughed and said, "You must be raving bonkers".

As with most trips it's always a case of check in and drop your luggage off and meet by the bar ASAP. The warm hotel bar was a blessing and I could dismiss the denim jacket straight away, but nobody told me I would need some arctic gloves to hold the beer!! Who on earth decided that serving lager in a frosted chilled glass, when it's -20 outside, was a good idea is beyond me? The option of a nice warm real ale was not available up here!

So, here I was 3,336 miles from home on my first works test trip. It was both exciting and daunting. There was the thrill of the adventure but also the worry of the task ahead and making sure you did a good job. I didn't really need to worry though because I had the best team around me who all had a lot of experience of cold climate testing and two of them would become my very close friends throughout my Jaguar days.

After a good night's sleep, I open the curtains to the sight of a very cold and snowy Main Street and what cars and trucks were passing left vapour trails in the winter air. I didn't really want to leave the comfort of my warm room, but needs must, and I put on my thermal underwear, a couple of layers of clothing and the only coat I had. It would be enough for the trip to work and I was told I could go shopping to get a better coat later in the day.

I met with the guys in reception and we headed down to the underground car park where our SUV had been plugged in all night to keep it warm. As the door opened and I took a good intake of cold air all my nose hairs froze instantly, and the strange sensation made me grip my nose straight away. This in turn pressed the frozen nose hairs into my skin like tiny little needles. The subsequent nosebleed lasted a second or two before that froze as well. Fletch confirmed that this is quite normal at these temps and the blocked nose of blood is referred to as Timmins snot!
All the newbies do this so it's good for a bit of banter on the drive to the workshop and I'm assured that there are a few more temperature related surprises to be had on the trip.

The workshop had been set up for the test teams to run cold climate durability of 17.5 thousand miles on each car. The cars would run on the road during the day and then come in for ramp checks and servicing when required. At night they would be parked outside for an overnight soak so that you could do a cold start in the morning. During the test period you would monitor every aspect of the vehicle and report back your daily findings to the Whitley Engineering Centre. We also had a reporting system to monitor faults called a CFR or Component Fault Report. These were raised and sent to the appropriate component owners who could then advise on a course of action to monitor or replace the part. With the nature of testing in Timmins the majority of issues were due to snow packing, ice build-up or things getting frozen and not working correctly. Our job though was to ensure the cars made it to the target mileage by the end of the trip in March so any faults had to be actioned very quickly and in cases where parts were needed urgently we get them flown out with an engineer on the next available plane.

On my trip we had the V12 engine being tested and the revised XJS which were both rear wheel drive and quite heavy so driving could be a little tricky for most newbies. I was fortunate to have been in Vehicle Proving and had some training on the wet grip circles at MIRA but even with that extra tuition we all had to do the ice lake driver training when we arrived on site. It sounded great and I couldn't wait to get going. We took three cars out to the lake and tentatively drove onto the ice and parked up to meet the instructors. As we exchanged initial pleasantries one of them said "Rule number one…never park the cars near each other as the weight will crack the ice" and sure enough you could hear it creaking under the combined mass of our three vehicles. They were moved really quickly!!!

It's the strangest feeling driving on the ice and it takes a while before you fully relax. I keep asking about the thickness of the ice and I'm told it's more than a foot thick and quite safe although there are a few things to be aware of. Having the sunroof open or a hammer in the car to break the glass is a must because if you go under when in the car the electric windows will not work. Never go off track to parts of the lake with snow on as the ice might be weaker or thinner in those areas. Never leave a car parked for more than two hours in the same spot and if possible, drill a hole next to the car and watch for water welling up, which is a sure sign that the ice is going down. One of the other key things was not to drive too close to the other cars on the lake as ice flexes and recovers after you have passed over it, so we are told to stay some twenty meters behind. Once you go through all of this it makes you even more nervous but after a few laps when you are drifting the cars like a seasoned rally driver you forget all about it and start to enjoy yourself. There was some additional reassurance though when they advised the team that they send out a special machine to measure the ice thickness every day and where it's a bit thin they drill an air hole and let the pocket of air out and allow it to freeze overnight.

After a full day on the ice I'm more than ready to head back safe in the knowledge that I can keep our very expensive prototypes in a straight line if I must although sideways is far more fun!

In my short time at Jaguar I had listened to a few stories about Timmins, but the main theme seemed to be about a place called the Mattagami or The Tog. From what I had picked up it was the centre of everything in Timmins if not the world. It was in essence a restaurant, bar & nightclub that just so happened to have a lap dancing bar as its main attraction that was not lost on all the visiting test teams that also included Land Rover & Aston Martin. Now I'm sure that most stories over the years grew to become folklore but they still made this Nuneaton lad a bit apprehensive. I had heard of them and knew there were some in London but not where I come from. So that night it was suggested we had a pizza and then move on to the Tog. Both Mark and Rob asked me if I was ok and I shrugged it off like I had been in a million times and tried to look as cool as a cucumber. Inside I was in fear of the unknown and I just followed the team into the darkness of the lounge.

The team selected a large table and started to order in the beers, but Mark pulled me to one side and said sit here with me and Rob as we can have some space. I wasn't sure why I needed space but let's go with it.

When the beers arrived, I just sat back and watched the proceedings. On the stage was a good-looking girl busting some moves to a great track by the scorpions and it was most impressive I have to say. Sitting right next to the stage were about three or four Canadian men who seemed quite keen to keep giving the girl dollar bills every couple of minutes for every good move she made. You could tell they were Canadian by the shirts, hats and the fact they were just not as reserved as the Brits. They got far too close and seemed too keen to stare which made it all a bit seedy and added to my apprehensiveness. Fletch asked if I wanted another drink and went off to the bar and returned with a round for the three of us and then a couple of minutes later one of the dancers came over and said "Hello David, I have come to dance for you", errrrrrr I think you have the wrong chap as I have not booked a dance! "This dance is courtesy of your friends Mark and Rob as a welcome to Timmins".

As a gentleman I can't possibly go into any of the details other than to say it was an eye opening four minutes of my life where I tried my hardest to maintain eye contact at all times but failed on a few occasions as she went through a full naked gymnastic routine whilst positioned on a small podium between my legs. Mark and Rob were grinning like Cheshire cats and the rest of the team on the next table were all giving it the thumbs up. When it was all over, she gave me a kiss on the cheek and wished me well for my stay in Timmins and if I wanted more of the same then just ask for Natalie. At this point I would like to point out to my children, who I hope will be reading this one day, that this all happened twelve years before I met your Mum!!

Well, it was certainly my first visit and it wasn't my last because there wasn't much else to do at night. Rob and I got to know Natalie really well over the four weeks and she liked us a lot. We were always very courteous and had a lot of chats about life. It turned out that she was a student from Montreal looking to make as much money as she could to pay for her fees and we both felt that we had to contribute to such a worthy cause so much in fact that I think Rob and I funded the rest of her course! On her last night she informed us that she was writing a book about her short time in the business and she was going to give us a mention. At the time Rob and I looked at each other thinking the same thing…I hope the book is only released in Canada! I never thought at the time that I would one day write a book and give her a mention!

I could write an entire book on stories from the Tog but I'm not going to, but I am going to give credit to those who featured in the Top five. Gary M, your ability to drift a remote control 4WD truck around the stage with a vibrator strapped to the top deserves a warm applause. Jim for managing to bite the thigh of a dancer without getting thrown out and banned from the club was a true reflection on you being the eldest person in the bar and you were a true gent that just had a blood rush in the heat of the moment. Steeley, you can't get away from being the only bloke who got up on stage to try and do a pole dance and failed miserably. I'm told the girls thought it was hilarious.

Gary Hone for marrying the owner of the Tog and never getting the teams any discount! And finally, to the Spam Gang, and you know who you are, for watching everybody else's dance without ever paying a dollar of your own money…shame on you!!

Outside of the Tog I must give a mention to my dear friend Paddy Healey who bless him could have a chapter all on his own. The Timmins Pizza place was the centre for one of Paddy's legendary moments and I wish I had been there to see it. At one of the team meals the guys all sat around a large table and everybody tucked into their pizza none more so than a rather hungry Paddy. On finishing off the last mouthful the team enquired as to how good it was. Paddy said it was fine if not just a bit chewy. They burst out laughing and showed Paddy that he had also ate the cardboard base that the pizza had been delivered on. You are a legend mate.

The trip I was on had been sent out for December as the first trip of the season and our job was to get the workshop set up and get the durability cars unloaded and up and running.
We had five prototypes running and each engineer looked after his own car. As well as running whole vehicle durability testing we had a folder full of test requests from engineering for specific vehicle tests like opening and
closing the doors during cold starts, check usage of windscreen wipers, monitor tyre temps for the whole test and record time taken to start the engine to name but a few.

One of the tests in my folder was to check the range of the remote-control lock/unlock key fob after an overnight soak and record a polar plot. What the engineer wanted was a measurement of the key fob from the car at various angles around it and various distances away to ensure it opened and locked the car. This seemed like an easy work instruction, so I set about getting this completed as my first task. I kitted up for the test and took a note pad and pencil to record the measurements. With the car already set up from the night before in the middle of our parking lot I went to open the car with the key fob. My massive thermal gloves made it tricky but doable, however I could not write on my pad with the gloves on and I removed one quickly and jotted down the data required.

In trying to put my glove back on I put the pencil in-between my lips to make it easy, As I went to take the pencil out though I found that the metal bit that holds the rubber in had frozen to my lips and I could not pull it away. I ran into the workshop to get some water on my lips and release the pencil, but it was too late, I had been spotted by the team and they found it quite amusing as always. I didn't mind at all because having a laugh and getting the work done made it a better trip.

Once all the cars were up and running and specific vehicle tests were logged in you could get quite warm and sweaty in the old corrugated hanger and that got us thinking one day. If you worked up a really good sweat what would happen if you ran outside and back without any winter gear on. Only one way to find out. The heating was turned onto max and we busied ourselves around the workshop for the morning until we all had a good sweat on. We then opened the door and made a run to the parking barrier and back. The temp recorded was -24°C and before you had reached the barrier you were frozen, and it felt weird. Not sure there were any lessons to be learnt from this scientific test, but it was a good lunchtime activity none the less.

With the cars up and running durability and with Mark and Ray at the helm Gary suggested that it might be a good idea to take me out and show me some of the routes and few places of interest. We kitted up for the drive like a couple of explorers because in this environment you never know what to expect. We had all our cold weather gear and, in the back, we had a full medical kit, sleeping bags and emergency food, torches and spades because if you get stuck in a snow drift out on route it may be a while before another car passes.

The roads are all covered in hard compacted snow, but all the vehicles have snow tyres which are very good. We head out into the wilderness that is Northern Ontario and make our way to a town called Cochrane where I see my first Polar Bear. Unfortunately, this one is a statue called Chimo and it's rather large. I take some pics and we head back onto the road north and then onto route 655 back towards Timmins. It's a great feeling being out in the wilderness with snow all around and just shooting the breeze with somebody. Gary is great company and has a lot of knowledge of the area from previous trips and we discuss many aspects of working for Jaguar. He is passionate about the company, the brand and the products and this enthusiasm is echoed in the vast majority of engineers that I worked with in that period. On the back of its recent success at Le Mans there was a wave of optimism in the company and with Ford taking over the future was looking ok.

Gary explained his expectations for the visiting engineers in that they should enjoy every moment and party hard if they wanted to, but they should always be in on time the next day and work even harder. I liked this view and it worked for me. Over the years I like to think that I had the same attitude for the trips that I lead overseas.

As we trundled back at 55mph Gary told me about some of the hazards you can encounter when driving. One year a car went off the road and into the ditch and before the next car came along it was completely covered in snow and the exit route was through the sunroof. The team in the next car could just see the hazard warning lights flashing in the snow. The biggest problem though was wildlife with Moose and Wolves. The Moose are huge!!! I managed to see a couple when out on the route and they were bigger than the truck we were in. Gary told me a story about a development trip the year before when they were doing XJS and on a night drive the car hit the Moose and it went in through the windscreen. It had been standing facing up the road, so it was hit from the rear and folded up with its backside going in through the glass first. Luckily both occupants were fine although they stank when the poor animal lost the contents of its stomach in the car. They said the stench was really bad!

A logging truck stopped to help and once they were all sorted the driver asked if he could have the Moose as it would make a lot of meals for the winter months. As the boys preferred MacDonald's to Moose they let him have it. They managed to throw it over the front of his truck where he strapped it down and drove off.

As we got closer to Timmins Gary suggested a stop point to take some pics in a wooded area and a chance to feel what it's like out in the wilderness. We stretched our legs and made our way into the forest and soon came across a kill. I think it must have been a small Moose and the wolves must have had a good meal out of it. On close inspection we decided it must have been a recent kill and we wondered if the wolves were still around.

Within three or four minutes there was a distinctive howl that could have only been a wolf which was soon followed by a couple more. We looked at each other, made some OMG facial expressions and legged it back to the truck and got out of the area as fast as we could. I never got to see a wolf on that trip but just the noise was enough.

We headed on back to base and I really appreciated the time out to talk and relax with the boss in our own little space bubble.

The workdays were long 10hr stints that followed with full on evenings of beer, food, more beer and the Tog for 6 days of the week that gave us Sunday off. Most engineers used this to get laundry done or just stay in bed and chill out for the day but with only four Sundays available I felt the need to get out and when Mark suggested a trip out to a Gold Mine, I was in. However, this trip to the mine was going to be in a helicopter over the Porcupine Mine.

The gold rush in Northern Ontario started around 1909 and peaked in the 1940's and 50's and continues till this day. We

47

head off to meet the helicopter and take to the skies which luckily are quite clear for a change. The snowy landscape is just amazing and looks like a giant white blanket over everything but as we head towards the mine it all changes and we are overhead of the biggest hole in the ground I have ever seen. I was expecting a typical mine with a small railway line going into a hole or something like that. I hadn't bargained for an open cast mine on this scale.

We circled around to take pics and then Mark told us that we were heading over to see the test base from the air. It only took a few minutes before we were skirting over the treetops to our workshop and there out front was an XJS parked up with Jaguar written in the snow. It looked great and the pilot hovered just above for more Kodak moments. As we turned to leave there was Ray Angliss waving at us from below. Mark had arranged this with Ray as he didn't want to fly and thought it would be something special for myself and Rob. It was a great way to spend some of your day off and far better than trying to go skiing at -30°C the following weekend.

The success of any overseas trip is down to the team you have and the team leader and this one was most definitely going to be a success although we did have the 'Spam Gang'. No names can be mentioned and to be fair the Spammies were all really nice chaps but collectively they were a bit like old mother hens and they arrived at their name because they didn't like to spend any money whereas most engineers spent all they had and more.

We would eat the best steaks and enjoy nice drinks, but the Spam Gang would have just a starter and a glass of water or maybe just one beer for the evening. Their motive was to save as much money as possible from the trip as well as making a nice earner on a 60hr week, but we hatched a plan that would dent the finances a little and make us howl with laughter. On one of the Tog nights the Spammies had done well watching the rest of the team spending a fair few dollars on dances and they mentioned that a couple of the girls were very nice so we had a cunning plan that we would pretend that these two girls were prepared to do a private Sunday show at the hotel for 240 dollars. That meant we all had to chip in thirty dollars apiece. There were grumblings from the spam gang, but they finally agreed that a relaxed Sunday with private dancing for thirty bucks was worth it. Having sown the prank seed Fletch was keen to have some extras just for the crack so we decided that this event should have some food as well and suggested to the team that another 5 dollars each was required for finger buffet food. Everyone paid up and all was going well.

Over the next day or so we then made believe that one of the heating engineers had not paid up yet but had also lost his wallet so we should all throw in additional funds to cover him until it was found. More groans from the spammies.

We arrived at D Day and agreed with the spammies to meet in Marks room, as he had a large area with a sofa and chairs, at 2pm and the dancers would arrive at 2:15. We gathered the team together and made our way to Marks room before 2pm, Mark left the room and hid out of the way. When the door knocked the three spammies were standing outside in their hotel robes and slippers and we were in jeans and T-Shirts. We howled and told them that this wasn't some sort of seedy Sunday session. It was going to be a relaxed afternoon for beers and lap dancing. Embarrassed they all flew back to change and made it back in five minutes. They all sat down and they told us how nervous and excited they were and then the door knocked.

Building up to the moment we opened the door and in sprang Fletch shouting "Surprise". So, it was at least 24hrs before the spammies saw the funny side. To this day I have no idea why they believed we would set this sort of thing up, but it was hilarious!

The start-up trip was a success, all the cars were running well and Ray had the workshop all sorted for the next trip out in January. All we had to do was shut down the facility, put the cars to bed and handover the keys to our Canadian landlord to pass on to the next team after Christmas.

As is the tradition for the last night we had a full team meal and then ensured we spent our remaining dollars at the Tog for dances and cold beer in freezing glasses.

On the Dash 8 back to Toronto Mark, Rob and I sat close together and discussed how good the trip was and how we had enjoyed the company of each other. I was staying on in Toronto for Christmas to stay with my great aunt, but we had time spare for a final meal that would be a fitting farewell for us.

As usual it was nothing but the best for Fletch and we had seats at the top of the CN tower. It was a lovely meal and everybody was relaxed and looking forward to going home. As we finished off, I sat for five minutes staring out at the skyline and it suddenly hit me as to what I had had just been through and just how good it was. I was in a privileged position to be testing prototype cars for the best car company in the world, being paid well, staying in nice hotels and having the best time with the best bunch of blokes I could have wished for.... even the spammies. What made it even better was I had established myself as part of the Vehicle Proving Team and forged a relationship with a cooling engineer called Rob who is still a great mate today and also found the closest of mates in Mark who became my best man and even though he is now retired some twenty seven years later I still hold him in the highest regard as a true friend.

The team departed and I made my way into Toronto to stay with Helen my great aunt. The plan was to stay over Christmas but I got all home sick on the second day and decided that I just had to fly home on Christmas Eve.

I managed to get my British Airways flights changed, said my goodbyes and headed off for the airport.

I was upgraded to club class upstairs thanks to the plane being almost empty and found myself all on my own. The lady from cabin crew was fantastic and we sat and discussed my Timmins experience in great detail. After a while I asked if it would be possible to go into the cabin (remember the old days pre 9/11). She went in and came out to tell me that they would call me in when we were cruising over the Atlantic.

I had a meal and true to their word I was asked to go in.

We had a great chat about everything to do with flying and the crew were also interested in my testing tales from Timmins. I sat there in the middle for about three hours and it really was the icing on a great test trip.

We landed on Christmas day and my poor old dad had to make the journey down to collect me in his in own 2.9 Jaguar XJ. He hated motorways and traffic so I really appreciated it. As we exchanged pleasantries he threw me the car keys and said "There you go Mr Jaguar, you take us home."

I never went back to Timmins and it has now closed as the company moved on to utilise other bases in Finland and Sweden that I also managed to visit, but I would like to go back one day, maybe in the summer to see what it's like under all that snow and ice. And who knows, maybe Shania Twain will be back home too.

# 4. Just How Hot Is Phoenix

During my Vehicle Proving days there was one place that I really liked the sound of and that was Phoenix Arizona. It was warm and sunny all year round and the city sounded fabulous. We had our own hot climate testing base in Dear Valley Park very similar to MIRA for its workshops but without the track. The base was used to support road durability and city cycles as part of the hot climate sign off as well as hosting hundreds of visiting engineers who wanted to do cooling, fuel, gearbox and aircon testing in the extreme heat. Phoenix in the summer months boasted 100deg heat and a trip up to Death Valley could see temps as high as 120degs if a test needed them that high.

The city also has a population of approx. 1.6M so it's about the 5th biggest in the USA.

My first chance to go there came out of the blue whilst I was out on a launch event in Montreal Canada in October 1995. I had been taking the press out in the new XJ when I got a call from Hoppy to say that there was a car with some issues at Phoenix and could I go down from Canada. It was a short trip via Chicago but the change in scenery and climate was obvious as we flew over Colorado and into Arizona. Lots of desert sand and clear blue skies all the way in.

The first thing you notice is that Phoenix is surrounded by mountains and is contained in sort of a bowl or valley which is why it retains so much of its heat and has little rain. The updraft of heat causes high pressure which keeps the clouds away and makes it great for testing. With subsequent visits over the years you can see that Phoenix has spread beyond the mountain ranges and some experts believe that by 2040 Phoenix and Tucson (which were 120 miles apart) could merge and how scary is that? Could you imagine Birmingham and London merging as one city due to expansion in the UK?

On one of my trips I had a great view of this expansion when I managed to get in on a mail flight between the two cities. One of our durability drivers was a pilot in his spare time and twice a week delivered mail between the two.

We set off from Dear Valley at about 6am when it was a little bit cooler and headed down to Tucson with a few bags of mail for company. We stayed quite low over Phoenix and the view through and over Squaw Peak was amazing.

Once out of the valley the forward view was just flat in all directions, so we climbed a little to put everything into perspective. With the extra height it wasn't long before we could see Tucson off the nose, but my view was taken up by something a bit closer. On our left wing was the biggest aircraft graveyard in the world Davis Monthan Air Force Base or as some refer to it as the boneyard. The US Military use it to store aircraft and there are approx. 4500 of them scattered around the desert floor. It looks amazing from the air and it's equally as impressive when on the ground.

In 2007 I had the pleasure of visiting the base and the PIMA Museum when myself Wayne Burgess (XF Design Manager) and Rob Till our South African photographer managed to blag our way onto the base for some PR shots with a mass of Hercules aircraft. I have never seen so many planes in one place and it takes your breath away.

Whilst there we went onto the Pima Air Museum for the most bizarre moment ever. We were just talking to the receptionist and a couple of the guys who worked at the base about how we were from Coventry England when one of them said "Wow, have you come to see your old girl land"? As it turned out an old Shackleton Bomber had been donated to Pima from air Atlantique in Coventry and was a plane I had seen a couple of times at Bagington a few years ago. We were taken onto the aircraft pan and watched the plane land and taxi in like we were some sort of VIP's. Without a doubt we were just in the right place at the right time.

Anyway, the plane back to Phoenix was now making its way down to Sky Harbour airport when the captain came over the loud speaker to announce that O J Simpson had been acquitted on two counts of murder. There was a round of applause and a few 'whoop whoops' from the passengers that made me raise an eyebrow. I guess the whole trial had captured the attention of the nation and they sure appeared to be behind their NFL hero.

As the plane landed and we made our way to the arrival hall and they were all still discussing the outcome of the trial. I collected my bags and made my way out in search of Margie from our base

As I weaved my way around the masses I could see a board reading David Moore Jaguar Cars and I made my way over to introduce myself. Before I could say hi Margie was already telling a TV film crew that this was the English guy who had just flown in. They thrust a microphone into my face and were keen to find out what the UK thought about the trial and its outcome?
Just for a laugh I asked them what trial they were talking about and they quickly told me it was O J Simpson.
I just could not resist and played dumb saying that I had no idea who he was and I'm sure the UK didn't either.
It was a short interview and they flew off to find another passenger.
I said hi to Margie and she asked me if I really didn't know who O J Simpson was? "Of course I do....his family run an orange juice business don't they?" Margie gave me the look and I just smirked.

When we got back to the base it was just great to put so many faces to the names that I had got to know over the last four years. The base was run by Hight Flexman and his resident engineers John Murchison and John Churchwood. They also had Margie as PA and technicians Vic Brown and Pete Perrone. As the base was set up for durability testing they also had a pool of test drivers that included Dan & Linda that became great friends during our high speed testing in Pecos.
Having had the tour to meet everyone I was keen to get to the hotel to freshen up and work out where to eat for the night and luckily Pete Perrone stepped up to the plate. Dave Steele, Mark Fletcher and Nick Gilkes had all got to know Pete over many trips and had already been in contact to say I was on my way over.
Pete was an absolute gent from day one and over the years we became the best of friends. We shared a lot of time together in Phoenix, Mexico and the UK and boy did we have a blast!

On the drive back to the hotel Pete invited me over to his place for dinner with the line "Mi casa es su casa", which is a Spanish expression meaning "My house is your house."

It was an offer I could not refuse and I have taken it up on every subsequent visit.

Pete's place became a second home to so many of the visiting engineers and his family loved having us round for what must have been a continual stream of BBQ's. His wife Chic was like our American mum on each trip and over the years we watched his two boys Alex and Jonathan grow up into adults.

The BBQ's became legendary as the food and hospitality were amazing. Pete had the biggest BBQ we had ever seen and boy could he cook!! You could easily put on 8lbs of weight on a short test trip to Phoenix.

Pete also had a fantastic swimming pool and the teams spent many happy hours just lounging around at the weekend like one of the family. On my first dip into the pool the strangest thing happened. The temperature must have been over 100 degrees and the pool was the best place to be. I remember just floating around enjoying the cold water and Pete shouted over that dinner was ready. I hopped out and the weirdest sensation happened. As I stood by the side of the pool I suddenly started shaking as if I was as cold as ice but within a couple of seconds I was toasty warm. The phenomenon is called 'evaporative cooling' which is something we are not used to in the UK.

I did a few trips to Phoenix and I have to say it was a place I could easily move to. Even without the stunning hospitality of the Perrone household, life in Phoenix was the business.

When the alarm went off every morning you woke up to the whirring sound of the aircon unit and the temperature of your room is just perfect. You get up and open the curtains confident in the knowledge that there will be clear blue skies and loads of sunshine. It's just the best feeling to start every day.

Breakfast was always a challenge trying to stick with cereals when the option of pancakes, sausage and bacon was giving you the hard Paddington bear stare.

Hopping into the car you then go into waftmatic mode, select some tunes and cruise into work without a care in the world.

Even with all the hot climate testing issues and loads of cars to work on nothing gets you down out there.

Our working days are based on ten hours a day to make sure we get the most out of the cars. Durability testing is a key part of the overall vehicle sign off and the cars that are subjected to 60,000 miles in Phoenix as well as 17,500 miles in Timmins at temps down as low as -50 with wind chill.

We ensured that the cars went to the extremes as we were a global company selling cars worldwide and we wanted to make sure ours were the best.

The extreme heat in Phoenix really gave the cars a hard time as they were subjected to some of the hottest temperatures known to man.

They would even conduct heat soaks in the compound on some days, just baking in the summer sun to see how the plastics, wood and paint would react. I remember going out into the compound to check the new interior of the XJ and by accident I lent my arm against a chrome door trim finisher and it felt like I had been branded with a hot poker.

The air conditioning units had to work the hardest though and the Heating Ventilation & Cooling teams spent weeks and months in Arizona testing to the limits. It's just so hot that on one of my trips the guys showed me how to fry an egg out here by cracking it on top of the bonnet of one of the hot soak cars. It fried within seconds!

Another unique Phoenix thing was sun tea. The guys at the base used to put tea bags into a large jar of water, and leave it outside in the summer sun, where it got nice and hot and ready for a cuppa. I was asked if we did this in the UK but I'm not sure we could even get the water hot enough if we left it out for the entire summer.

At the end of every shift the test cars would return to base and the data cans were downloaded so that the data could be interrogated by the test teams to ensure that everything was working as it should. Occasionally the cooling teams would take the cars off to Death Valley where it was even hotter just to push the boundaries a bit. The heat in the valley was just like being in a fan assisted oven and you had to make sure you had plenty of water in the car and some good emergency back up plans should you break down out there. I remember being told that a chap that did break down in the valley and was stranded ended up drinking the fluid out of his car battery. He remembered that he topped it up with distilled water and thought it would be fine. You laugh at the stupidity but then wonder what sort of heat induced turmoil he must have been going through to get to that point. During one fuel systems trip we had to stop in the desert to do a cold fuel fill from our refrigerated truck and I have to say it was seriously hot. We made head scarfs from spare T Shirts tied around our heads and covered ourselves in factor 50 sun cream everywhere else. Luckily, we all climbed into the cooled fuel truck most of the time to chill out.

The test that we were conducting though was to see how the fuel reacted and if it did what happened in the fuel pumps and fuel lines and the result on engine performance and emissions through vaporisation.

A lot of people always ask if the extreme heat would cause fuel tanks to explode but the truth is that petrol would take hundreds of degrees of heat for that to happen that have never been registered on earth!

The extreme heat does affect aircraft though and one year there were about fifty flights that had to remain grounded at Sky Harbour airport as a result of high temps (116°F) which makes the air too thin to take off safely.

On many occasions, during four or five weeks of continual heat that you don't ever experience in the UK, I wonder why they don't ever have a drought. And where does all the water come from? There must be lots because there are around 185 golf courses in Phoenix, and they all have lush green grass which certainly does not come from the average annual rainfall of about 7". Well the answer is that it gets pumped in from Lake Mead and its source of the Colorado River some 300 miles away but it's not an endless supply and scientists predict that it could run out by about 2023 when the levels at Lake Mead drop so low that it can't be pumped out which is referred to as 'Dead Pool'.

Obviously, the heat will continue so the city has already built underground reservoirs to store water and are even contemplating pumping water in from the great lakes some 1700miles away!

Another observation from my days in the sun is that nobody walks anywhere in Phoenix, even if it's just a short walk to some shops you drive. You get up in the morning to chilled air from the air conditioning units, which stay on all night as a result of summer night time temps of nearly 100°F, you get into your car quickly and get the air con switched on to max chill and then drive to the mall and park as close as you can to the doors so that you have the minimum distance to air conditioned shops. Sometimes the odd visiting engineer would wander out for a local stroll, but they only ever did it once. If the heat hasn't made them go very red then the continual horn blowing and finger pointing from the locals would make you go red anyway.

With a permanent base out in Phoenix people get used to seeing test cars driving around in various types of camouflage and after a few years it's quite normal and the locals don't take too much notice. The press on the other hand do. They know our routes around the city and they also know we will be out at Death Valley during the peak temperatures of the summer. They will even base themselves at the same hotels in their quest to get some scoop photographs of our prototypes.
It's a good earner too and I have heard of prices going into the thousands of pounds for that perfect picture.
On the normal durability routes it becomes a bit of fun to try and loose them and get away but some of them try really hard to stay with you. In the more recent time with the advent of mobile phones you only have to pull into a petrol station and everybody is taking pictures that end up on all the automotive blogs.
During one trip through the desert our transmission team stopped to get fuel and decided that the camouflage needed some repair work. They replaced the tape and threw the old stuff in the bin. A couple of weeks later the old camo tape was

for sale on ebay because it came off an XF prototype test car. People are so strange!

Camouflage has come on leaps and bounds over the years from the bolt on panels of XK8 development and then the bagged S-Types in 1998. We now use some of the most advanced swirly wraps that are designed to fool the camera lens and prevent people from seeing the detail of the car. Keeping it on the cars in the extreme heat is another matter though.

So in answer to the chapter's question 'Just how hot is Phoenix', well it's damn hot!!!

# 5. Texas and the road to hell

The heat in the desert summer sun is scorching hot and I'm standing in a dry gulch beside a test track in the middle of Texas watching the nodding donkey's pumping oil from below the earth's crust. Oil that will ultimately be refined into the petrol that I need to run my Jaguar test cars. As the headlamps peer out from the heat haze in the distance you can just about see the front end silhouette of the new XJ Supercharged Cat but in no time at all it roars past and disappears into the distance but the music in my head begins to play and I hear the dulcet tones of Chris Rea "This aint no technological breakdown, oh no, this is the road to hell"

High speed durability is a major test for any Jaguar. Our cars are sold in Germany and on the autobahns, customers can drive them to the max as they have no speed limits. Our road load data teams had spent a lot of time collecting data from the German Autobahns and developed a high-speed cycle to replicate those conditions. For many years the company used a famous test track in Italy called Pista di prova di Nardo Della Fiat or just NARDO for short. The track was built in 1975 as a Fiat test track and was a 7.8 miles long ring and was a perfect location for our high-speed testing. However, when Ford took over there was a requirement to complete joint testing and therefore a suitable test track was required in the USA. We sent out a team to look at Michigan and also Fort Stockton, but they came back with a very secure location just North East of a place called Pecos. The firestone test track was 9 miles around at 2700ft elevation making it bigger than Nardo and perhaps the biggest circle track on the earth.

Pecos itself is located between Dallas and El Paso on the I20 with a population of approx. 9000. The town is the regional centre for ranching, oil and gas production and is well known for the cultivation of cantaloupes and famous for its claims to be the place of the world's first rodeo in 1883. More recently, between 1995 to 1998, it became the home of visiting Jaguar Test Engineers.

After the first couple of test trips the teams were complaining that there was not much to do at night, the next nearest town was 100 miles away and the facilities at both the track and the hotel were poor. My then manager Dave Hopkin called me into his office for a chat and asked me if I would like to be the next team leader to Pecos? He wanted me to go with an open mind and see what the issues were and come up with a plan to improve them. I was delighted to be asked and started to plan for my trip. I gathered as much info as I could about the area and got the team together at MIRA for a pre-test brief to warm them up for the trip. As well as visiting test engineers I had my own mechanics from MIRA and also Paul Merriman from Powertrain to keep a check on the engine development.

We set off from a very green UK heading west to a place called Midland Odessa in the heart of Texas. As we descended from 37000ft it became very clear that the colour green had vanished without a trace to be replaced with various shades of brown and sand and the landscape was covered in circles with black dots in them as far as the eye could see. We landed and I was met by my old mate from MIRA Nick 'Mad Dog' Gilkes who was wearing a very fancy Stetson. 'Welcome to Texas Dudes'. It was so hot just walking to the car and we were so happy to get the aircon on for the 100-mile drive to Pecos.

We get onto the I20 and I think that was the last time Nick turned the steering wheel for the next hour and a half. It was the straightest road I had ever seen, and I would be seeing it many more times over the next few years. As we trundled along, I realised that the circles and black dots were actually oil wells. Hundreds and thousands of them just bobbing up and down pumping up litre after litre of oil. The smell was overpowering even with the aircon on re-circ. We eventually rolled into town and dispatched our bags to the rooms and headed for the hotel watering hole by the name of the Purple Sage. It was your typical western saloon type bar and I loved it. Nice bar, a few tables, a juke box and a pool table so what more could you ask for. The bar tenders were already friends of the outgoing team and they welcomed us in like brothers. I had a good feeling already that this was going to be better than I thought.

The next morning after a hearty breakfast burrito and large coffee we head on up to the test track where Nick will show us around for a couple of days as we complete a team handover of the site and we take control of the testing so the others can fly back home.

Normal test trips last between four to six weeks so teams are always keen to get home and let you get on with it but with Nick here we all wanted to make the most of it with all the boys. The technicians head into the garage and Nick and I do a tour of the test site and meet the hosts. The track was mainly being used for tyre testing and run by the Smithers Corporation, but the main workforce were predominately Mexicans.

Having exchanged pleasantries, I was keen to get a look of the track so Nick and I got suited up and climbed into one of the XJ prototypes. Heading out from the hut we made our way out onto a clockwise direction today and accelerated past 100mph in no time at all. The track had three lanes of tarmac with the inner lane for slow speed testing and the outer for high speed. We moved into lane three and commenced a high-speed cycle that would see us complete the nine miles in just over four minutes. If it wasn't for the track control building you would not have realised that you had just completed a full circle because the road looked straight. This place was enormous. It was so big that the USAF used it as a waypoint for the B1 planes flying out of Albuquerque. On many fast laps I would often see the shape of the B1's coming from miles away and would adjust my speed so that our paths crossed at the right point. I would start flashing my headlights and the planes would always shake their wings in acknowledgement which was quite cool in a Top Gun sort of way.

On my first test laps I noticed a lot of sand swirling in from the right which Nick informs me is a dirt devil or mini tornado which can affect the car at high speed. As it's coming in from the right, I move towards it and sit to the right of the middle lane. When it flashes across the track the car instantly pulls to the left going with the devil and we must move over by about six feet. It's something you don't get at MIRA that's for sure and although the first one makes your backside grip the seat after a few you just get used to it and it's not the worst thing that can happen on the track. Even in this barren dusty landscape covered in dry tumbleweed there is wildlife a plenty and the one you don't want to see at 155mph is the Javelina Pig or Wild Boar. These solid animals just run in from nowhere and things can get quite messy as one of our drivers found out when he hit one and wrote the car off. The impact pushed the bonnet up against the windscreen and totalled the front end of the car. As part of our high-speed driving course for the track we are advised to just hit things if they run out at you as swerving to miss them you could end up rolling the car. Our other problem was snakes and in particular the rattlesnake. They would often slither onto the test track to soak up the warmth off the tarmac and although they were no threat to us in the car, they brought other unwanted attention in the shape of Vultures and Buzzards. If we hit a snake these massive birds would come down from high to snap up an easy meal. On very hot days with no wind they struggled to get airborne again but had worked out that if they spread out their wings, which are about 2m wide, they could use the air turbulence from a passing Jaguar to get them off the ground. It only happened to me once and it was quite daunting I have to say.

Having had an action packed day, we all climb aboard the hire cars and head back to the hotel and straight to the Purple Sage for a few cold ones and a round of pool or two. The harmony and balance of a test team is important to me and I always ensure that we do as much as a group and generate that team spirit that makes a long test trip much more fun. I had been on other tests as a support engineer and seen the group just break away and end up at loggerheads causing an atmosphere that ruins the trip. With the feedback I had had from Hoppy I was not going to let this happen on my shift. With that in mind I left the bar for a while and set off to introduce myself to the manager of the hotel. I can't recall her name but she was a lovely older lady who had been in the business for years and I told her all about myself and the team and explained what we would be doing and I expressed a wish to get involved more in the community over the next six weeks. She smiled and said that she had seen the other couple of test teams but they had not really made contact with her, but she would love to get us involved. As it happened they were having a special evening for two old ladies to raise money for their hospital treatments and she asked if we would like to attend as guests. I accepted straight away and made my way back to the boys to interrupt the beer and pool. I gave them the low down regarding the evening which wasn't what they were expecting but explained my logic to them and requested a minimum of one hour of their time to support this good cause. With a few mumbles and a bit of banter about having sun stroke I corralled them into the events hall attached to the hotel.

The place was heaving, and we were shown to our table and the evening kicked off with a speech from the hotel manager (let's just call her Patsy) to welcome everyone and run through how the money helped these two dear old ladies for the year ahead. Patsy then finished by announcing some special guests from England were here and would I like to say a few words. I stood up and thanked them for the invite and explained that we were there for six weeks, we always miss home and home cooking and we would love to get involved with the community and get to understand their life and culture. We received a nice warm round of applause and we all sat back down to tuck into some food. As the evening moved on they announced that the next event would be an auction for some homemade treats and I said to Nick that I really liked the look of a massive Pecan Pie that was up for offers and suggested that's the one we bid on. I told the team to all throw in ten dollars each and explained that Nick and I would control the bidding as we really wanted this pie.

When the moment came, we were ready. The bidding started at a couple of dollars which we raised to five. The floor raised it to seven dollars and Nick jumped in to ten dollars. It took some coaching from the chair to get it to fifteen, but I think local knowledge had already set its limit on this pie, but they had not bargained for the Jag Lads to be there. I said to Nick lets blow the bank and just bid against each other as it was for a good cause. I raised the bid to twenty dollars and there were already rumblings in the hall. Nick was very dramatic and raised it to twenty-five. As we had eighty dollars this was going to be fun and at one stage Nick bid fifty dollars and then bid against himself at fifty-five dollars and the locals were loving it. We milked the occasion and finally secured the pie for the full eighty dollars. The place erupted and we had a standing ovation much to the embarrassment of the rest of the team. We all went over to hug the old ladies and they were lovely and thanked us for our contribution. Having made it to a minimum of an hour we released the team back to the bar, but Nick and I had some social networking to do and stayed on to be treated to some good old country dancing, free beer and excellent Texan hospitality.

Back at the track the next day we were back in the groove checking the test cars and logging up the high-speed miles when the phone rang. It was Patsy from the hotel asking what time we would be getting back and could I see her in the office when I returned. My first thought was 'Oh no what's happened? One of the team must have got into trouble last night at the bar or something. We arrived back and I went over to the office. Patsy had a grin on her face and thanked me again for the kind gesture the previous evening. 'Well David, we have a problem, you see, in your speech you mentioned being away from home and you miss the homemade food' well we did so what's the problem? 'Come this way young man' I followed Patsy into the kitchen, and I could not believe my eyes. On every available surface there was food laid out like a feast for a king. There must have been twenty or so pies. There were pecan pies, blueberry, cherry, key lime, cheesecake and apple. There were also cookies, muffins, pancakes and biscuits. 'So, what would you like to do with them David?' I sure don't think we can eat them all that's for sure. It was hilarious to see, and I was moved by it. I decided that we would take a few different items to work but the rest could go into the hotel restaurant with donations to go to the charity. We really pigged out the next day that's for sure and to this day I have not tasted finer homemade pies!

With the outgoing team due to leave the next day the team leaders decided that it was best if we stayed in that night and

65

had a good old session in the Purple Sage and a pool tournament. What locals were in the bar were invited to join us as long as they were in for the duration. In my team I had two drivers sent in from our Phoenix test base by the names of Dan and Linda and as Native Americans they were loving our work and play ethics. Even in just two days I knew we were in for a great trip with these two in attendance. Dan loved his pool and proceeded to advise us on their rules and the tournament began. What we hadn't bargained for was the appearance of some strangers to the bar and all of them were women. They ordered drinks and came straight over to the table and placed some quarters onto the rail and asked to be part of the game. Well the more the merrier I reckon, and I head off to the bar for another round and I'm introduced to the head of the group. I explain who we are and where we are from and then she advises me that they are a travelling, all female pool team. Oh…just Oh! After the first few frames, we have hardly had a shot and they are taking us apart. Nick takes me to one side, and we decide that the only way to get out of this humiliation is beer…lots of beer! Now I can't go into details of how the game de-railed, but we managed to make some sort of recovery before we ended up having drinking contests at the bar with tequilas. All I can say is that I'm so glad I was not flying home tomorrow (sorry Nick). The evening was a hit for all the team and the girls behind the bar asked if we could do this every night? Err I don't think so.

It was a quiet drive to the airport the next day as the outgoing team slept the whole way. We said our goodbyes and Nick wished me all the best for our trip. As we left the airport and headed back to the I20 there was only one thing to do…. get that Chris Rea CD on as I need the 'Road to Hell' track for the long drive back. The drive was not too bad if you had company to talk to but with two cars to get back Paul had to drive the other one and we had left Dan and Linda at the hotel to recover. Anyway, I always found good loud music the best way to stay alert but that was not always the case. On a couple of late night missions into Odessa the long drive back in the dark was difficult and you actually get road hypnosis, because all you can see in the desert is the white lines in the road to focus on. We had one incident where the driver did briefly fall asleep and we went off the road. It was a shocker for sure and after that we all decided to swap the driving around, play loud music all the time and purchase a box of caffeine tablets just to be sure.

I really enjoyed being a team leader on these trips and it made me feel valued. You had £250,000 prototype vehicles to test, important data to collect, reports to write, timelines to meet and the responsibility of ensuring your team are safe and well looked after. You are also judged on your performance and rated in your end of year assessment so my view was always to look after the team, and they will look after you. It's a privilege that Jaguar put you in this position and they trust you to do a great job and I think that I revelled in it and the trips made me who I am today.

Our day job for the next six weeks was to sign off the new XJ Saloon and make sure the engine and gearbox changes were robust. Like most Jaguars I had tested so far the engines were bullet proof and the mileage accumulation was easy. Dan and Linda were loving it and the only thing Paul was struggling with was the snakes and creepy crawlies and if there was ever a chance to capture something to show Paul, I was straight in there. Our two regulars into the workshop were tarantulas and scorpions and I often caught them in big jars to show Paul and watch him run a mile. The rattlesnake incident took it to a new level though and I mean a new level quiet literally. We had this visitor slide into the shop and with the help of some very experienced Mexicans the snake was caught and dropped into a large bucket for us to view. I got Paul to take a look and he nearly passed out, but I told him to stop being a wuss and let's get our cameras to take a few pics. Having taken a few close shots with my camera Paul did a few shots from a distance but I needed something special. I found a stick and told Paul to prod the fella and let's see him sit up and rattle his tail. The Mexicans just stood quietly observing our strange ways and it's a good job they did. I set up the camera and Paul gave the chap a good prod, but I did not expect what happened next. It shot straight up in the air like an un-coiled spring. I dropped the camera and flew backwards a few feet but one of the Mexicans stepped up and pressed the stick onto the head of the snake and picked it up. I looked around but could not see Paul anywhere until I heard a tiny voice calling down from above 'Is it safe yet?' he had only shot up the drainpipe and was sitting on top of the office block. The poor lad had played his hand now so it was not going to be the end of creepy crawly surprises for him then which leads me nicely into the Texan Two Stepper. Whilst having a mooch around I came across what Dan told me was a horny back toad. They look weird but are quite harmless. Well Dan and I decided to rename it the Texan Two Stepper on the basis that if it bites you only take two steps before you die. Armed with this false info I take the toad into the workshop to show Paul and tell him about its deadly bite. If that wasn't enough to take him over the edge you should have seen his face when I picked it up by its horns. Poor lad was going bonkers at me and we couldn't keep it up any longer. Linda and Dan burst out laughing first and then I did. Paul didn't instantly find it funny, but he warmed to it over the trip and by the end was keen to tell the story himself.

Dan and Linda also got me good and proper one day when a snake about 7ft long slithered into the workshop and I almost had a baby. Paul by this time was already in the next state and nowhere to be found. Dan made me stand dead still as the threat was massive from this one and you just don't muck about when you are told that. As he manoeuvred himself around the snake Linda stopped it from going under the workbench and I was thinking 'My god these two are brave souls'. When a gap presented itself Dan shouted 'RUN' and I rocketed out of the shop like Linford Christie. Whilst still hyper ventilating I noticed Dan walking out holding the snake and he said 'wanna stroke him? He's harmless really' Touché Mon Amis!!

Well the snake stories don't stop there. On one of my trips I had the pleasure of working with a MIRA technician by the name of Mark 'The Tatts' Wallace. He was a gentle giant of a man really, but I would never want to cross him. We were having a great trip and nothing seemed to faze him much apart from rattlesnakes. As fate would have it, we had one enter the shop and one of the local Mexicans dealt with it quickly and I was presented with the rattle from it later. Out in Texas the snakes are considered vermin and even the police are called out to deal with them. I dried out the tail in the sun and as it rattled really well I decided to put it on the end of a strong piece of wire and have what I thought would be some fun. Mark was installing a roll cage into a car in the workshop so I sneaked up on him and got the wire into the car and shook it. Oh my god could that massive lad move when he needed to. I very briefly stood up with the rattle on a stick until the full force of 'The Tatts' hit me like an articulated truck. The wall of the workshop prevented him from taking me into the desert and I just dangled like a rag doll for a second or two. He did drop me in the end and gave me that look and had a deep breath. 'Well youth, you had better sleep with one eye open tonight' were his parting words as he left to make a brew. In hindsight it was a daft thing to do on the basis that these snakes are deadly serious, but Mark eventually saw the funny side and we had a good giggle about it later with the team.

Marks revenge though came quickly. On arriving at the workshop I decided to start the day off with a trip to the toilet which had just been built for us with a nice changing room and shower facility. As I was sitting there reading the paper I could see something coming under the door in my peripheral vision. It was the biggest black snake ever and I leapt off the throne and dived into the shower cubicle clutching the shower curtain. My panic was short lived as the giggles and laughter from the other side of the door gave it away. It was nice work by Mark using some rubber hose and a bit of paint. We called it a truce after that though!

Our only other venomous creature in the shop was a scorpion which was easy to catch in a jar and easy to take away from the shop and let it go but we decided to take a couple of pictures first. The jar was placed on a car and we went to get the cameras. On returning the jar was knocked over and Sid the Scorpion ran off though the vents just under the windscreen. We all looked at each other in a WTF moment before the driver on the next shift said 'I aint driving that car till you find it'. Five hours of careful teardown were required before we extracted the little fella and set him free. The locals told us to just throw it out the back but us English guys need to make sure it wasn't coming back soon and drove it a mile or so from the base before we let it go.

We did see lots of black widows and tarantulas, but they never gave us any issues at all. The one thing that did cause us an enormous amount of problems though was a mouse. One of our test cars had been having some electrical issues and my technician Steve Coleman could not find the problem. We kept checking the car which would run ok for a while and then just do all sorts of odd things. After a couple of days of investigation Steve said, 'take a look at this' and there in the electrical harness to the fuse box were a load of chewed wires. As we were sorting out the mess the mouse popped out and then ran off into the engine bay. For an hour or so we tried to catch him much to the amusement of Mexican workers from their test shop. They stood and watched the two crazy Englishmen for ages as we tried all the humane ways to catch it. The biggest of the spectators stepped up and pushed us away. He picked up a screwdriver and when the mouse popped out it was all over in a flash. He passed us the screwdriver with the deceased mouse on the end and left. I looked at Coleman and I think we both had a tear in our eyes.

Over the years I really enjoyed being in Pecos and the townsfolk enjoyed us being there too. After the charity auction success, I did a few talks with the chamber of commerce, the police, the fire station, local schools and the hospital and we were getting invited out on all sorts of trips. We went horse riding in the desert to round up some cattle, went to a bull ride where they told me I could not have a go and attended some fantastic BBQ's. At the hotel they could not do enough for us and the girls at the bar Michelle Bell and Toyah Walker always had the drinks ready when we got back after a long day and made sure we were invited to all sorts of family gatherings. Dave Hopkin was pleased that we had turned it around after the first couple of trips and now the teams were looking forward to the place. I have always had the view that when you do an overseas assignment you have to get involved with the community and don't be shy.

The first test teams were right though, there were not a lot of places to go and it didn't have the attractions and nightlife of Phoenix but if they had just gone out and engaged with the locals there were loads of things to do. I was talking to Toyah one evening about taking a drive out to some caverns and she said that we had to go there at a specific time but would not tell me why. She picked us up and we drove out for an hour to an amazing setting. It was typical western stuff with cactus's, dry riverbeds and some caverns that you could imagine cowboys or Indians living in. Toyah had prepared us some food and told us to just sit down on the rug and enjoy the view. After a few minutes we were joined by many others all doing the same. There was a lovely warm breeze and you could hear the crickets chirping away just as you would see it in the movies.

As it drew darker fireflies were dancing around like crazy and it really had that western feel. People then started getting up and we joined them by the edge of a large cavern or sink hole as it might have been. Toyah said, 'be quiet and watch'. After a couple of minutes there was a strange noise and then from nowhere thousands of bats came flying out of the cavern. The sky was full of them for ages. I was blown away by it and so pleased to have been invited to see it.

As well as all this wildlife there were also several characters in this town and none more so than our local tyre store manager.. Nick had made friends with this chap and he joined us at the bar for beer and pool on a few occasions. He had that proper cowboy look and a big truck to go with it. We got talking one day about Texas and the gun laws and he was keen to tell us about their constitutional right to bear arms and all that stuff and he was surprised that we had nothing like it in the UK and even more surprised that we had not even fired a gun. That was soon resolved though with a trip to his ranch where we ended up shooting at tin cans and bottles until our arms could not hold up the guns any longer. After the shooting we met his family who ran the ranch and they were lovely. We spent a couple of hours talking about our life in the UK and what it was like testing cars for Jaguar which was so different to their very relaxed and

71

simple stress-free life. After a fantastic meal he took us back to the hotel for some more beer and a round of pool but as was often the case his pager went off and he had to head out on into the night to fix a tyre on a truck. On one of my return trips I went off to see him but was told that he unfortunately had to leave the state after some Federal intervention. I was then informed that the pager calls late at night were to pick up drug drops and pass them on to the network of delivery trucks running through the state and in exchange for his assistance he was given the option of a new life. It was certainly a very serious thing and I'm glad that we knew nothing about it until after the event, but there were certainly loads of other characters that made us laugh though and will be remembered fondly, like the World famous or maybe I should re-phrase that, Pecos's famous Billy Keeble. Billy was a country and western singer at the bar at weekends and the lads thought he was great. Billy put a cowboy boot on a table and all his tips were thrown in the boot. He was without a doubt a legend in his own mind and he got a real kick out of doing some very dodgy duets with the Jag lads. Top of the bill was my now gone feral mate' 'Mad Dog' Gilkes'. Nick stood up one evening for the finest rendition of 'All My Ex's Live in Texas' and it brought the house down!

Our usage of the facility continued for high speed sign off on the XJ V8 and then the new S-Types that we did alongside the DEW98 or Lincoln programme from Ford USA. This joint venture was an interesting time for us as it pitched our project test teams with those from Dearborn and sometimes the only thing in common was that we both spoke English…. well….sort of! It was fascinating to see how our work cultures differed so much. They were full of process, getting authorisation for everything and taking twice as long to do anything. The decision making was made based on their career aspirations and ensuring that it was all choreographed to a CV that would pull them through the system to management and beyond. Our way was based on doing the right thing for the product and the customer. We made fast decisions in most cases and reviewed them later. If they were wrong, we changed them. We also had more empowerment to make on site testing decisions without having to contact our managers every five minutes and that helped a lot for both the projects and the individuals. I remember one case of trying to cool a differential. The priority was getting the engine signed off, so we just needed to keep going and not worry about the diff at this stage. The guys from Dearborn spent hours on the phone and online chasing engine, axle and cooling development teams and their managers to get a group decision on what to do whilst in the same time we had completed two or three iterations of handmade cooling scoops and the car was back up and running doing mileage accumulation. I came across this on many occasions and discussed it at various levels and I hope that both teams learnt something from each approach.

Humour was another thing that we had no common ground to work with. The American humour was lost on us most of the time and they had no idea of how to deal with banter which we had in bucket loads. If we made a mistake in the workshop or dropped the proverbial bollock, then you could guarantee that there would be a lot of mickey taking from that moment onwards and they found it hard to deal with. Looking back, I think it was because they never looked at the team as a group of mates. They were always work colleagues and nothing more. On one of my joint trips I decided that we should have a full team meal at the finest establishment west of the Pecos River by the name of 'The Jersey Lilly'. This was a dry watering hole where you had to take your own beer and they cooked the steaks. So, the priority for the Brits was to order as quickly as possible with no faffing around at all. Stick to the basics and get the beers open unlike our partners who dissected the menu, wanted to know how every steak was cooked, needed extra sides, different toppings and a pre ordered box to go. For one night only everybody was getting a T Bone steak and fries end of.

Somewhat shocked that our quick-fire decision making in the workshop also carried over into the restaurant they started to ask who wanted chilled glasses for the beer but the look from my team gave them the answer….no glasses required here! After demolishing the dinosaur sized steaks the waitress asked us if we would like anything else and she was not ready for the answer…. yes….another round of steaks please which nobody had ever asked for before. When we finished round two even the chef came out to say Wow. They were so impressed that they only charged us half price for the second round and that my American friends Is how we rock n roll in our team. To end the proceedings, we decided that what better way to capture this occasion than a Kodak moment and this was a pre-selfie period where you handed your camera to somebody else and asked them to take a photo for you. I had often admired the horse riding saddles on each booth and asked if we could all sit on those. 'Well sure you can, let me fetch some hats and ropes too' she said. We climbed up for a great pic and the locals in the restaurant loved it, even when Paddy Healey (who is scared of spiders) put his hand in a web and when told of this nearly passed out and fell off the saddle and landed on a table sending food everywhere. Suddenly there was spontaneous laughter and we had found the perfect level of humour for our colonial cousins!

With our test base firmly established we started doing more testing and the cars were getting better and faster and I was delighted to be taking a team out to test the first Supercharged XJ Saloon. It was great to drive and the noise from the supercharger made it sound really powerful. We started off the test with caution and then just got faster and faster. It's a beast of a car and loved by all the drivers especially Linda and Dan from Phoenix. They both used to come in after Vmax laps with huge smiles and couldn't wait to get back out. I had warmed to them both especially Linda as she was our only lady driver. In fact, Dan had warmed to her even more and eventually went on to marry her when they were back in Phoenix.

Anyway, on one of these glorious days I dispatched Linda onto the track for a Vmax run and waited for her to return for a ramp check but she didn't arrive as planned. I started to worry straight away and had already got into a spare car with Paul Merriman when an alarm sounded, and we could see smoke in the distance. I flew straight out onto the track and buried the throttle to the floor leaving the track control trucks way behind. As we pulled up by the smoke, we could see that the car was on fire and we ran through the sand and tumble weed to see if Linda was still in it. She wasn't and we panicked for a second until we heard her calling us from the other side of the track. Having established that she was ok she told us that there were too many snakes in there, so she decided to sit far away on the tarmac. We grabbed the fire extinguisher and put out the fire quite quickly which had been started as a result of a very hot exhaust igniting the dry tumbleweed. Linda had told us that the car had an issue and she let it ride off the track and into the sand.

Whilst we were talking the trucks arrived and then suddenly we realised that the car was back on fire again. We had used up all the extinguishers, but the track team told me a fire engine was on its way. Paul and I were handed some fire beaters and told to go out and stop the fire from spreading. My only concern was to try and save the car, but I was advised otherwise by the track team. Paul's concern however was that Linda had just said that there were loads of snakes out there and he was not moving anywhere. The head honcho thrust the beater back onto his hand and said 'listen up, do you really think snakes are going to stay for a BBQ'. We both started beating out the flames, but it was a losing battle.

The fire truck finally came around the track. It was probably a late 1940's post war thing that looked like it should be in a museum or set on fire itself. It overshot the fire by quite a way and had to reverse back. I later found out that the brakes were shot and needed replacing! The fire crew got the hose out and I stood back to let the professionals take over. There was a lot of revving of the engine to build up some pressure I guess and then the hose guy signalled that he was ready and pulled the lever. Well, what happened next was like a moment off an old black and white comedy film. The hose guy fell flat on his face as there was no water pressure at all, in fact there was no water at all. Lots of shaking of fists followed and the head of the track started hollering at the fire crew who were being sent back to fill up with water. In the aftermath of the fire they found that the truck had a leak, but it was never checked to see that it was full of water. Just a bit of an error I reckon!

The comedy hadn't finished though as the fire crew went to drive back around the track but were stopped and told to go down the service road which was a dust track. As they sped off the truck came to a shuddering halt and the back lifted in the air. We ran over to find that the propshaft had broken and dug into the ground. You couldn't write this stuff. The manager of the base told us that we should leave and go back to the hotel and they would deal with it. They already had fire trucks heading in from Odessa as the fire was by now out of control as the tumbleweed was burning and rolling in the breeze igniting everything in its path. When we got back to the hotel we decided that a cold beer was needed after being in that heat, so we all sat down in the bar. The girls were concerned as they had heard about the fire and the local crews were already digging ditches and doing controlled burns to protect the town. It was all very serious. Michelle put on the telly and we could not believe our eyes as there was already a news helicopter in the sky reporting on this massive fire spreading for miles … Ooooops !!

The next day I attended a number of safety meetings and enforced quite a few changes to their operational procedures. As Jaguar was the only proper high-speed user our concerns were listened to and in fairness actions were taken quickly. Rapid response vehicles were put in place with more fire extinguishers and the fire truck was repaired and a daily check sheet was created and signed by each crew to ensure that it had water and worked.

Although we could see the funny side to this, after the event, doubts were raised about us using the facility and those doubts were confirmed when a couple of years later when in the space of two weeks there were two testing accidents at the track resulting in the tragic death of both test drivers. With great sadness we had to pull out from the test track and we never returned.

Pecos was however a great place full of friendly people living a good life and I loved the place. I got to enjoy country music, barn dancing, horse riding and generally trying to fit in and be a

Texan. I ate the biggest steaks, drank some good beer and enjoyed the company of everyone I met so thank you Pecos for making me an honorary Texan!

# 6. Track Driving Tales

I have to say that over the years track driving has to be one of my favourite activities and looking back I wish I had taken it up as a sport. I dabbled with bike racing and loved it and in some of my activities through Jaguar I think I could have made a good job of it in a car, obviously not on the scale of a Lewis Hamilton or a Colin McRae but maybe just hold my own. As it is, I didn't become a racing driver but I did manage to get onto various tracks driving lots of great cars so I'm very lucky and very happy. I had some fantastic mentors and great tuition along the way within Jaguar and also had the honour of taking in some coaching and tips from some experts including Jackie Stewart (F1 World Champion), Colin McRae (World Rally Car Champion), Johnny Herbert (F1 Driver), Andy Wallace (Le Mans Winner), Paul Gentilozzi (American GT Champion) and Chris Harris (Driving Journalist & Top Gear Presenter) to name but a few.

I spent a couple of occasions with Jackie Stewart and you can see why he was world champion. At MIRA I did some work with him on the Dunlop circuit and he was silky smooth. He used to put an apple on the dash and get you to drive around the circuit without moving the apple. When I tried it, I was lucky that the apple stayed in the car but you listen and keep trying and eventually you get close.

With Colin McRae it was a whole different ball game. I was out on the Dunlop circuit doing some demonstration laps in the XJ Supercharged car and sharing the time slots with Colin. During one session he was watching from the barrier and when I pulled off he came over and was really interested in the car. I told him all about it and he asked if he could have a go. Whilst I checked with PR, he did some hot laps in his Subaru and when he came back I had clearance to let him drive. We climbed aboard and he did a few installation laps and pulled up. He was smiling and asked how on earth I got a car that was so heavy to race around the track. I think I shrugged my shoulders and said, "Lots of practice". He then took it back out at pace and within two or three laps was sliding it around like he had been in the car a hundred times. He was an absolute gent and was good enough to answer all my questions and even told me that I could drive his car later in the week but unfortunately it didn't happen as he had to fly off to Corsica. He was a legend and a massive loss to motorsport but some years later at Brands Hatch I did get a go in his Focus!

Although I had many drives in many different cars the following are just some short anecdotes from each track that I have had the pleasure of driving on in some of my favourite cars.

## MIRA
Having had a few years at MIRA around all the circuits it difficult to pick out my favourite. I had some great fun in lots of competitor cars such as the Dodge Viper, Porsche 911 GT, Corvette and McLaren and hours and hours of fun in all the Jaguar products but the best of them all by far was the XJ220. It was the wall poster car in my bedroom and the car I salivated over at the Birmingham Motor Show and inspired me along with the XJR9 to join Jaguar. All through its development I used to dream about driving it. After all, it was the world's fastest sports car capable of 220mph.

My chance came when an event was being planned at MIRA to take clients and some VIP guests out around the high-speed track and having done some work with PR I was asked if I would help co-ordinate the event and look after the car. The driver for the event was going to be Andy Wallace who drove the XJR9 to victory at Le Mans in 1988.

The car arrived and I looked after it like a mother hen and it received so much attention. I managed to drive it up to get some fuel but would have to wait for the event to see if I could have a go.

I met with Andy on the test day before the event and he was brilliant and very relaxed. I warmed the car up with some care on the No 2 Circuit and it felt so good even at reduced speeds. When Andy was ready, we swapped over and made our way over to the No 1 High Speed Circuit. He did two laps to settle in and warm up and then completed a couple of flying laps that planted me right back in the seat.

At the time it was mind blowing and I held on for dear life having never lapped this fast. All my dreams had come true and I was smiling from ear to ear. Andy was happy and decided that we should do some runs and drive the car like he would on the event day and see how hot it got. We went back out and during a full throttle acceleration there was a bang and the car came to a halt. Unfortunately, there was a problem with the gearbox and that was the end of that. I didn't get to do a flying lap on my own but who cares, I had just done a few laps with a great Le Mans Winner who was much faster than me anyway!

I did manage to get a drive a few years later though at road speeds and that was very special to me as well as taking my boss Dave Hopkin for a drive around Gaydon, so dreams do come true!

## Gaydon

The engineering centre at Gaydon became my new home in 2009 when I started to work for ETO (Engineered to Order). We were developing the specialist vehicles for JLR and I enjoyed testing the XKRS, XFRS and Project 7's on the circuit, which actually started its life as an RAF base for the V Bombers, but even with all this exotic hardware it was a 1960's car that stole my heart.

My boss in ETO Kev Riches had left us to join up with the Heritage Team to recreate the Lightweight E-Type race cars and build up a new division called JLR Classics. He always kept in touch and I badgered him a few times about getting into some of the Classics so it was no surprise that he called one day to say they were shaking down the Lightweight and if I wanted to meet up at the track I could have a go. Well, chances like this don't come around too often so I dropped everything and headed up to track control.

Kev had already done some test laps in the car, so it was ready for me to have a go. With the car valued at over a million pounds I was a tad nervous, but I put my crash helmet on and buckled in.

As with any drive you always take your time on the way out to get a feel for it and then step it up a bit as you gain confidence but with little time available it was a warm up lap and a flying lap and back in.

I was surprised at just how light it was compared to the modern-day sports cars and that helped it feel really fast off the line. The really large steering wheel was like something out of a bus and there was no precision at all so you had to manage lots of input, but it felt alive. The noise from the powertrain and the sound from road surface all came into the cabin and added to the experience and I was not at full throttle yet. When I did ask it for the beans it was just beautiful. We absolutely flew up the main straight and I kept it to about 130mph, but it could have gone much further. I stayed on the side of caution because if something went wrong you don't want to be in the history books as the man at the wheel when a million-pound car got written off like the late Norman Dewis did with the XJ13 at MIRA.

On the in lap I could not contain my excitement and there were smiles all over the place. This was a piece of Jaguar rich racing heritage and here was I driving it. Another great day on planet earth!

**Fen End**
This track was also an airfield many years ago and home to Hurricanes, Spitfires and Mosquito's and was called RAF Honiley. The site closed in 1958 and was converted to a test track and became the new home of Lucas Girling and then Prodrive under the leadership of Dave Richards who also had the BAR F1 Race Team. It was taken over by JLR in 2014 and gave the Special Operations Division their own track.

As part of SVO I did a few drives there with F-Type SVR but the drive that stands out is the final shakedown and sign off of my first JLR Classics Vehicle called Project Nicko. This car was a bespoke XJ6 Series 3 for the Iron Maiden drummer Nicko McBrain.

Nicko had bought the car from new when the band first became famous and he left it in the UK with a mate and it sat unused for many years. He decided he would like a full renovation done with some extras and I had the task of doing it.

Due to the timing and needing to get the car out to the USA we didn't have any time to do the mileage in the car that we wanted so we arranged with Nicko to take him to Fen End and sign the car off with him on the track.

This wasn't a set your hair on fire moment due to lots of power and thrills, but it did make the hairs on the back of your neck go up. I just didn't know what to expect and I was worried about what Nicko would think having waited a number of years for this moment, but I needn't have worried. The old girl purred, and the ride was so lush that he raved about it. We did about ten laps with Nicko and Wayne Burgess, who designed all body modifications, and when we pulled up there were lots of hugs and a small amount of dust in the eyes. I was thrilled for Nicko and so proud of what we had all achieved.

We didn't break any lap records or trip out any of the sound meters but it was definitely Rock n Roll!!

## Millbrook

Well here we have a proving ground that was not an airfield in its previous life.
Millbrook Test Centre was opened in 1970 by Bedford and then passed to Vauxhall but is open for specific vehicle tests and events to anybody.

Its two key features are a 2-mile banked bowl for high speed testing and the Hill Route with gradients and jumps that have been seen on Top Gear and also one of the James Bond films.
The bowl and hill routes are both exciting and for me I had to do a launch video for my XKRS-GT here and that was quite special.

82

On the hill route we pushed the car hard whilst also having a chase car filming the event. I have done this before on the road at slower speeds but doing it at speed was something else. We had radio contact with the film car, and we got constant updates from the producer telling us to pull left, pull right, accelerate, slow down, let us pass and then full throttle around us. Having had a blast on the hill route the team wanted some high-speed action on the bowl again with the tracking team in another car. We spent quite some time just drafting the film car and were then told to drop right back and give it all the beans and fly past the camera making lots of noise. With the GT this was not a problem and the final result on video looked absolutely epic. Feel free to check it out sometime on YouTube under The New Jaguar XKRS-GT. The video still makes me tingle watching it so just think what it's like in the car. It's a dirty job I know but somebody has to do it.

## Smithers Texas

In the previous chapter Road to Hell I talked about my time at Pecos where we were doing high speed sign off at the Smithers Test Track. It was an enormous circle 7.8miles long and you could keep your foot to the floor until the fuel ran out. I love doing high speed testing but at this place you had to be on guard. There was the extreme heat, wild animals and freak weather conditions to throw in the mix and when you are sitting at 155mph for 7.8 miles with your foot planted to the floor you need to be focused.

For me the best drive I had was in the new XJ Supercharger when the project manager told me to run with an unlocked calibration and see what the max speed was. I completed a number of warm up laps and far more ramp checks to look at tyres, fluids and torque check marks before I was happy to go. I entered the track and the weather was perfect for the run. I let the car get up to about 140mph and waited till I was near the control tower and then I let it go. To get the max speed I had to run with aircon off and leave the front windows down a bit to get some fresh air in. Sweat was running down my face inside my full-face helmet and I was trying to relax my grip on the wheel as the car edged up to 180mph. At this speed you are covering 264 feet per second which when you think about it is a long way.

If anything happened at that speed it would take you about the length of three football pitches before you came to rest. You never think about this at the time but always work it out after the event.

I only did the one flying lap to record the data, but I had just become the first person to drive the fastest ever Jaguar production car. The occasion was toasted at the Purple Sage with one or two beers that night.

## Blainville Test Track Quebec Canada

In 1995 I was invited to go out to Montreal by John Webb who was the head of Jaguar Canada to run a Ride and Drive event for all the dealers to experience the new XJ Saloon.

We were based at the Blainville test facility in Quebec that had about 25km of test tracks but the one we were using was the high-speed oval. It was 4 miles long and the banked curves at either end were at an angle of 46° which gave a neutral speed just over 100mph. This means at this speed you can take your hands of the steering wheel and the car will go all the way around the bend itself.

For the event I was joined by the late Craig Hill who was an ex Canadian Sports Car Champion and Jacques Bienvenue who drove for Porsche at Le Mans. Both had retired from racing but they sure could still drive fast.

Whilst doing our set up and test drives before the event they taught me how to draft other cars as they do in NASCAR racing. We were out on the track with 5 cars and they put me at the front and said go for it. After a couple of laps one then two cars went past me, but they were the same spec and had the same power. They then got me to run at the back and stay close to the last car. I did this and found that at the max speed I was not using all the throttle pedal because the draft from the other 4 cars was pulling me around. When we got onto the straight you floored it and pulled out which slingshots you past the others. It was bonkers.

We had so much fun showing off the cars to all the dealers and they in turn loved it. For me the highlight was on the last day when we had some journalists in the car and I had to do an interview whilst driving on the oval. I hit the banked curve at 111mph (neutral speed) hit the cruise control and took my hands off the wheel. I turned around briefly and said "so what's you next question" both nearly passed out.

## Arizona Proving Ground
Situated not far from our Phoenix operation the Arizona Proving Ground was opened in 1985 by Volvo and then sold to Ford in 2009. It had a 2-mile oval track with banked curves and a 2 mile straight road and some other small tracks. It's right in the middle of the desert with hardly any rainfall and low humidity. Its normal summer temps are 100°F so it's very hot!
All my driving here was done in XK8's during its test and development and all I can remember is the heat haze when you were going down the 2-mile straight and lots of sand and cacti.

## Argeplog & Kiruna Sweden
Test tracks with a difference. No tarmac, freezing cold weather and lots of snow. The tracks are based on the top of frozen lakes and perfect for our traction control teams during the winter months. The Kiruna location is now our winter base for cold climate testing. I had a blast being asked to try our new Traction Control Systems in Argeplog but even more fun in Kiruna when asked to support a PR Film for the launch of XF. The production team from Tangerine Films were keen to shoot the XF going through its paces on the lake doing slaloms through cones and drifting it around large corners and creating a lot of snow dust.

Being filmed is not everybody's cup of tea because it takes a lot of time, you end up repeating many shots for better angles or better light and then you end up sitting around a lot too but I enjoyed it so much. Maybe I missed my true vocation in life and I should have been an actor?

One of the highlights after having thrown it around the lake for a day was to get off the lake and do road shots of the car going through a pile of snow creating an impressive snow wake behind the car. I had studded tyres fitted and went for it. I hit the snow pile at 120mph on top of compacted snow and the end result was fabulous.

Also on that trip I made friends with a cameraman called Casper Leaver who did all the filming for Top Gear and although we had so much fun driving and filming the funniest part of the trip was a visit to the ice hotel where Casper licked an ice sculpture and obviously got his tongue stuck. We thought about leaving him there for the night but ended up pouring some warm water over his tongue to get it off.

We ended the night and filming with some vodka in ice goblets at £10 a shot…never going there again!

### Lara/You Yangs Australia
The Lara test track has to be one of the prettiest places to go. It's situated west of Melbourne and approx. 26 miles from Geelong and sits in Victoria's Western Basalt Plains Grassy Woodlands. It has about 80km of test tracks and was opened in 1965 and is home to Ford's test and development programme in Australia and what makes this place is that you have to watch out for the eastern grey kangaroos.

I had the pleasure of setting up two S-Type cars for the most epic road trip in the history of everness. There was nothing too special about the tracks or the set-up drives but what I remember most was meeting my test team here for the first time before we set off on a 10,000 mile adventure which you can read about in chapter 12.

### Mallory Park
This place is my home track and I raced here on my uncle's Yamaha TZ. It's quite a small track but has the famous Gerrard's Bend and my favourite corner The Devil's Elbow. I didn't get to drive a Jaguar here, but it was Jaguar related. To enable you to drive at the Goodwood Festival of Speed you had

to have an MSA racing licence and to get that you had to complete an ARDS (Association of Racing Driver Schools) test at Mallory Park.

I was despatched to Mallory and completed the course in a Renault Clio Cup Car. I was expecting to do a lot of laps but with my track driving experience and knowledge of Mallory we only did a couple of laps before the instructor said, "I don't think we need to do any more of this you're fine". It was a shame because it was great fun and the car was a little rocket ship. It would be good to go back and race there one day on a bike but I'm not sure Mrs Moore would sign that one off.

**Castle Coombe**
Another perimeter track of an old WWII airfield that opened its doors to racing in 1950.

I had been asked by the Sales and Marketing Team to take down my XKRS-GT and do some demo hot laps for the crowds at a race day and also to meet some customers who had paid deposits to get one of the limited edition vehicles. Mark Twomey and I, who was also from SVO, headed down to the track and it gave Mark a chance to drive the car on the way down for all his help on the project.
The GT was a phenomenal car and I will spend a lot of time telling you how we developed it later in chapter 17.

With a break in racing I hopped into the driver's seat with a potential customer in the passenger seat and set off for a couple of fast laps. A bit like driving at Goodwood for the first time you are very aware of the crowds and the cameras ready to capture anything and everything. It's a nervy,time but you do put it out of your head when you get going and concentration takes over. The car squats down under full load and the track tyres grip the surface better than anything I had tested before. When we exit the track, the customer is really pleased and can't wait to get his deposit down.
When we get to the pits we stand there talking but keep getting interrupted by a chap who is a bit keen to talk to me. I advise him to wait five minutes and when I'm done, I will see him. He

throws a complete tantrum and storms off telling me he wants his deposit back. I chase after him and calm him down a bit, but he is spitting feathers about not getting a drive in the car. I'm taken by surprise because I was the only one driving today.

It turned out to be poor communications as the dealer had told him he could drive the car instead of just a go in the car. In order to turn things around quickly I invited him to a behind the scenes tour of Gaydon and then a go in the GT. He was much happier now and when he did come to the track he was thrilled to bits. He bought one of the limited edition GT's and sent me a few shots of it outside his villa in Portugal looking splendid.

## Silverstone

The home of motorsport and a place I remember well from the Kenny Roberts & Barry Sheene duals in the 500cc World Championship and the Epic Nigel Mansell win in F1. I had my own little F1 moment in 2000 when I was asked to support a magazine article about our new S-Type Vs the F1 car. It involved a lot of driving in the S-Type and my only go in the F1 car was being towed behind a camera car but even that was so cool. I also did some great laps in the XKRS and got a passenger lap in the most gorgeous car on the planet the CX75 but my most memorable time was through our events team when I was asked along to support an event with HSBC where all the motor manufacturers took their latest vehicles so that all the fleet rental companies could try all the products in one day. I had the very latest S-Type R at the time and had a pit garage along with a standard S-Type and an XJ. The plan was that I would take people out for a hot lap and then they could try the other cars themselves with an instructor in the passenger seat.

I had a walk down the pits to look at all the other hot lap cars and realised that all the drivers were from motorsport. Lots of touring car drivers and ex F1 drivers with BMW, Mercedes, Audi & Peugeot. Well this could be fun!

I spoke with a couple of the race drivers who told me that all the customer drivers would indicate to get out of the way, and we could then focus on a hot lap. Gentleman racing or what?

I belted up and had my first victim. She was a nice lady who had never been on the track before. I talked her through what we were going to do and she was well up for it. As I sat in the pit lane waiting for the lights to go green I slipped in some Queen on the CD player for my favourite track 'One Vision' to play. As we were waiting the slow build up to the track was adding to the occasion and just as Brian May hit the power chords the lights went green and we flew out of the pits as if it had all been rehearsed. My passenger was firmly planted back in the seat and I had to check she was ok. A thumbs up and big smiles was all I needed for the out lap. By the time we came around for the hot lap there were about four or five hot lap cars together and I tagged along for the most exhilarating lap ever. In terms of hardware I had an advantage over the other cars with my supercharged engine, so I managed to pass two of them on the straight but as I braked into Copse they shot past and I was chasing again. Through Maggots, Becketts and Chapel we were almost touching bumpers and I was screaming with excitement as much as my passenger. On Hanger straight the S-Type R breezed past the Vauxhall, Volvo and BMW but into Stowe we were all in a line as if it was the last lap of a grand prix. Into Vale I lost my line a little and sat behind the gaggle of hot lap cars as we whistled past the customer drives who did keep out of the way. The next batch of corners came and went, and I found myself approaching Woodcote so I gave the S-Type all it had and managed to pip the Volvo and BMW as we went over the line and backed off for a cooling lap.

I had always wanted to try my hand at this and here I was mixing it with Touring Car drivers at Silverstone. It was brilliant and I was lucky to be in such a powerful car as I'm sure on equal terms they would have left me behind but who cares, for that one lap I was there and another box ticked on the bucket list.

As we pulled into the garage I looked at my passenger, who had stopped screaming by now, and asked her what she thought? "Well apart from the mild heart attack, almost peeing myself and having the shakes it was F in brilliant young man". She got out of the car and was telling everybody how good it was and how we went past some of the race drivers, the noise was epic and the handling was so amazing. You could not ask for anything better than making somebody happy and taking them out of their comfort zone for a while.

I repeated this for three years in a row and must have done nearly 200 laps in all and hopefully left about sixty people with a good story to tell at the pub. Job done I reckon.

### Brands Hatch

At the end of the S-Type R Launch the PR Team were concerned that the car was getting some bad press in some of the car magazine blogs so they decided to approach Piston Heads about a track day they were holding at Brands Hatch and agreed a deal for us to attend and let some of the drivers have a passenger lap.

With my previous exploits at Silverstone I was asked to support and I was hardly going to say no!

I was positioned in pit garage no 1 and sitting in my empty car I watched about 20 cars file out onto the track. It was a mixture of Subaru's, Mitsubishi's, Hot Hatches and the odd Lotus and Porsche. They were all keen to get out and flex their muscles. As I was sitting there one of the organisers said, "Why don't you just go out and have some fun".


I put the helmet on, strapped in and made my way onto the circuit and joined in the fun. Within a couple of laps, I was passing everything that I came up to and giving one or two punters a bit of a surprise. After a lot of fun I decided to pull in and cool off a bit and grab a drink. When I returned to the garage we suddenly had a few customers appear for a fast lap or two and through the day we had a lot more. The general consensus of opinion was that the car was awesome, and they couldn't wait to tell their dads about it. The proof was in the pudding because the next blog had them raving about the S-Type R and how it passed their mates on the track.

You can't help but smile at these events really. People fall in love with your car and they go away smiling! The bonus for me on this event was that the Colin McRae Rally Focus was there, and I finally managed to get a brief go in it.


**Oulton Park**

Another great PR day out as a result of Chris Harris wanting to have a go in my XKRS-GT.

The car had grabbed some great headlines from the static shows but now it had to prove itself at the hands of this great car journalist come race driver.

I had met Chris a couple of times at Goodwood and he was also my instructor at the sideways challenge event and I rated him highly. It was a cold damp October morning when we rolled up to the pit garage and handed over our one and only prototype into the hands of a legend. It was a very damp day but that didn't stop him from having a ball on the track sliding it around all over the place. The track did end up drying out and his opinion was that it was a weapon!!

Why not watch the video online? Just type in XKRS-GT Chris Harris and enjoy.

I managed quite a few laps myself but although I was there to look after the GT it will always be remembered for a discussion I ended up having with an old friend Don Law about an idea we both thought was worth pursuing and I left the track with my head spinning and excited about a skunk project which you can read about later in Chapter 19 'Above The Clouds at Pikes Peak'.


**Cadwell Park**

A great track that I learnt to race on with an RC30 Motorbike and also did my High-Speed Performance assessment with John Lyons. Managed to drive the track in the opposite direction in an XJS

**Elvington Runway**
I'm afraid I can't give away too much on this one other than to say myself and Dave Warner had a very special XFR and a chance at breaking a record or two so we both went up to this long runway and did some epic runs but unfortunately the project got canned and we used the car for something else. I have to say Mr Warner did a brilliant job with the engine and its calibrations and it would have been a monster but that's just the way things go sometimes.

**Coltishall Runway**
RAF Coltishall was awesome. My best mate Mark Ashfield who I met at RAF Lossiemouth in Scotland was now based there and I spent many weekends at the base looking, sitting and then flying in Jaguar jets.

Over the years the RAF had borrowed some of our cars for various events and I even did the odd chauffeuring job for them, but the best times were when they did their own air shows for friends and families. On one of these shows I managed to get the XK8 Bond car and the F1 Car to put on static show and I also had an XKR to drive on the runway as part of the show. We decided that it would be good for the spectators to see a car versus plane race so Wing Commander Bob Judson and I worked out some details, and the race was on.

On show day I had to go to the event brief with all the other pilots and I even got my own call sign X-ray Kilo 8. We lined up next to each other on the runway and the plan was for the plane to just do a normal full power take off and I would do a full throttle acceleration up to a finish line someway down the runway. As I glanced over at the pilot Toby Craig I floored the Jag and gave him "The Bird" as I blasted away. This was a mistake because he lit up the reheat and was soon making up ground on me.

The Jag had pulled out quite a lead but as I approached the finish line there was an almighty roar as the GR4 jet screamed over my car at about 50 ft. pulled back on the stick and went ballistic. The jet wash was amazing, and the car got blown around a bit but it was all under control. I did a power slide and brought it back to centre runway just in time for Toby to do another low-level pass over the car that was captured by the RAF photographer. Well….that's something you don't get to do every day!

## Conningsby Runway

When the RAF Jaguars were being decommissioned they moved to Conningsby for their final 6 months or so and I was lucky enough to spend a week or so there when we helped paint a jet (another story for later).

They were holding a 6-squadron day and I was invited to take a car up to put on show and as always at the end of the event I did some fast runs on the runway to show the team just how good an XKR was. It was during these runs that one of the pilots told me to cut across a taxi way and take a short cut back to the hanger but this was not liked by the control tower and before I knew it I was being pulled up by the Military Police. He was about to give me a lecture through the window and I assume some sort of civilian court martial when he noticed the number of stripes on my passengers flight jacket. It was nice being called sir after that and I'm glad the pilot was next to me!

## Cosford Runway

The Cosford Air Show is a massive yearly event and I was asked to support this with an F-Type and the James Bond car to set up a stand to promote STEM which stands for Science, Technology, Engineering & Mathematics. We obviously did the engineering bit and they then had other STEM ambassadors for Science with Professor Brian Cox and Maths with Carol Vorderman.

As part of the event I was asked if I could take Carol around our stand while she was interviewed on local radio. She came over and the introductions were made but then an old mate from RAF Coltishall 6 squadron Graham Duff interrupted and told Carol that she shouldn't go out on the track with me under any circumstances. I laughed and carried on with the talk and caught up with him later. Graham had progressed to the Red Arrows after the Jaguars and I asked him why he was with Carol. He told me he had flown her in for the day in his own plane. "Well that's a bit special from the Reds isn't it? Is that what you do now…taxi rides" "No Dave, Carol and I are an item". Errrrrrrrrrrrrrrrrrrrrrrrrrr "Well I never, you're batting above your average there". I had to check hello! magazine later and he wasn't kidding.

After the event I raffled off a ride in the Bond Car for charity and took a young RAF cadet for a spin. As a result I was then offered a return to Cosford for a private tour the following year where I managed to do some F-Type demo runs on the runway for the technicians which is always a pleasure.

Having finished this chapter it's made me realise just how lucky I have been to visit these places and meet so many people across so many walks of life. In doing it we all have one thing in common which is our love of cars and at every event you have to say the car is always the star!

# 7. Mexican Madness

In the run up to the launch of the all new S-Type we had to do some altitude tests to ensure that all the powertrain calibrations were working correctly. With my experience of overseas testing Vyrn and Pete were both keen for me to head up this trip and I have to say I was quite excited. Mexico was somewhere on my list to visit for its culture, its food and some great all-inclusive holiday packages but then I received the security brief from Ford, and I thought 'hang on a minute, this could get quite serious'.

There was lots of advice and many horror stories from previous trips to read and it made you sit up and take notice quickly. Jaguar had also conducted some durability testing in Mexico City and the team had their own stories including being held at gunpoint by the drugs squad on suspicion of being drugs barons after a local shop keeper reported two black Jaguar Cars and eight wealthy looking men going into a restaurant and also local police stopping our test cars and demanding cash payments for minor driving issues. The biggest horror story I read in the brief was of a man who had his drinks spiked and woke up in a bath of iced water with a letter on the wall telling him to call the hospital ASAP as one of his kidneys had been removed. I'm not sure if this was correct or not but it did send shivers up my spine and I was a tad nervous to say the least.

Another report on the list regarded laptop security at the airport, as a few engineers had their computers stolen just on the other side of the scanning/security search area. Mexico airport at the time had a strange system of traffic lights in security where after you put your bags through you pressed a button and waited until it went either green or red. This supposedly random process meant that you got searched on a red but ok to go through on a green. I was a little suspicious on the random thing as I'm sure everybody was being watched or scanned and they only made it red on those that looked dodgy. It's a shame they didn't have better security on the other side of the x-ray machine. As laptops went through they got lifted by the thieves really quickly. The advice we were given was to send somebody though first without a laptop to wait and check others through.

Armed with pages full of security advice, several travel injections and the assurance that Ford Mexico would provide us with full security I boarded the plane ready for the next adventure but not ready for the people I was sitting next to. I managed to get the front row of seats with extra legroom, but I was sharing it with a Mexican family returning home with their young child. Now normally this would not be an issue as I'm quite sociable and I do like kids but this one was starting to smell really bad. I think this poor little thing had been subject to some chilli, refried beans and guacamole and the air was thick of its fermented after tones. I expected the parents to change the little one as soon as the seat belt warning lights had gone off but oh no. They sat oblivious to the stomach curdling stench but what do you do? I considered a direct approach of 'excuse me your baby stinks, can you change it' through to a more reserved approach of 'oh, has he got a poorly tummy'. I opted for a safe retreat to the hostess area to get a drink and some fresh air where I also spilt the Mexican Beans on my stinky neighbours. The cabin crew were very helpful and took control of operation clean up. When I returned to my seat the poor little thing was now crying its eyes out as a result of being cleaned up by its stressed parents. Well please note my choice of words of 'poor little thing' a couple of times because after about three hours of crying and chilli infested flatulence my patience was wearing just a bit thin. I plugged in the earphones which muffled the noise down a bit but not the smell. During those hours I actually considered opening the doors and drop kicking the little rugby ball out of the aircraft. I mentioned this to my new Cabin Crew friends who must have been horrified at my suggestion because within ten minutes Stinky and its parents were moved to another area of the plane and my space bubble was a better place for it. This episode has actually affected me now that I have children of my own and I would be mortified if I was in the same situation and would hate it if my seat neighbours were thinking the same about my kids as I did about Stinky.

I mentioned this to Steeley some years later after he took his tribe out on holiday but he advised me that when they are your kids balling their eyes out and are causing chaos you just don't care about anybody else and you certainly don't worry. They were wise words but even now many years later I still can't entertain the idea of my children on a plane until they are at least ten years old.

Now this whole episode was nothing compared to what some of my engineering friends have encountered. Steve Botterill was on a flight to Canada that had to turn around and come back to Scotland. When it was just short of half way a passenger got up and tried to open the door (maybe he was preparing for his own Stinky to exit the plane). Another moment was when John Barnes was on a fuel systems trip to the USA which managed to get a depressurisation issue and he spent some time with the overhead oxygen mask in place and then there was a whole test team stuck out in Canada when the Gulf War 2 kicked off in 2003 and flights were suspended for a while. Apart from my snowy flight into Timmins and a scary flight out of Houston on a DC10 in the middle of a thunderstorm I have travelled all over the place with Jaguar without a hitch and I have to say it's a pleasure. I always had a sense of pride telling the booking desk that I was from Jaguar Cars. In fact, it often had its benefits especially with British Airways. I always dressed smartly and was always on my 'A' game with politeness at the check-in counters. I would always ask if they needed any upgrades and even if the response was negative I would leave them with either a Jaguar pen or lapel pin to say thank you.

I received two or three upgrades at check-in but also two upgrades after I had checked in being called back to say that a space had been made available. Going from the poverty seats at the back into club class was a big thing for the long-haul flights and it gave you massive bragging rights when you got to your destination. In all my flights I never got a first-class upgrade even when booked into Club Class but I did eventually taste the better champagne with first class tickets to Australia which you will see later in chapter 12.

The descent into Mexico City was everything I expected. At 10,000ft it looked like a heaving metropolis and had a thick layer of sandy brown smog just hanging over it. I negotiated the security traffic light system with a bright green light and met up with our Ford hosts for the trip who along with their own security men escorted us to our hotel. As we made very slow progress through the city you realise that every other car is a VW Beetle. There must be thousands of them in every state of repair. The locals call them Belly Buttons on the basis that everyone has one! As I'm gazing out of the window the driver tells me that on certain days in the city only even number plates are allowed in and then odd numbers on the other days which rotate all year and he also told us that the rich people have two plates and swap them around or have two cars with odd and even plates on them. I also noticed several crosses by the side of the road where people had been killed, a few dead dogs and even a dead horse. What a place!!

I was joined on the trip by my technician from Phoenix Pete Perrone and by some of the Mexican Ford Test Team to assist with logistics and security. It was great having Pete along from our Phoenix base as he was a great all-rounder for a trip like this and he also spoke the language. The reason for our testing was because the local Jaguar dealership was concerned about the off the line performance of the car at altitude. Mexico City sits at 7200ft above sea level and the thin air makes it difficult for the car to breath and therefore reduces its performance a little. You can't get better than real world testing, so we headed out into the city to complete some traffic cycles. It was clear that the car was struggling a bit and even the VW Beetles were getting off from the traffic lights a little bit quicker than us over the first few yards. To get some objective measurements we took the cars to a small test track and did some standing starts from 0-60mph. The data didn't look too great compared to our UK tests so we decided that we should run the cars down to Acapulco and complete some driving at sea level and repeat the standing start tests. The run down to Acapulco was amazing as we were on a sort of toll road that only the wealthy drivers used, and it was empty. We stretched the legs of the S-Type and it felt good at high speed which pleased the Mexican team.

We only had a very brief overnight stay in Acapulco but Pete did get us to take a dip in the Pacific Ocean, but it was far too brief as somebody mentioned sharks and we were out and dressed by the bar in seconds. We had a very early start and completed the performance tests at sea level and the car returned the data we were expecting so it was another rapid drive back to Mexico City where I wrote up the test report to send back to Whitley detailing our findings. We said goodbye to Mexico and headed back to the UK via Phoenix. I didn't think I would be coming back but the test report changed that!

S-Type development was going well, and we decided that to support the Mexican market we would change the ratio of the rear differential so that the car would feel a bit livelier off the line. The programme was keen for us to make the changes and send a car back to repeat the tests and suggested that I should go back too for continuity and I also had all the visa's and injections. My technician for the trip was my old mate from MIRA Steve Coleman and we were both looking forward to it. As we prepared for the trip PR contacted me to say that CAR magazine would like to join us to do a testing story and they thought Mexico would be perfect. A journalist by the name of Richard Bremner would fly out with us along with a photographer Jonathan Glynn-Smith. Jonathan was mainly a fashion photographer and the magazine wanted some different pictures from the normal car shots. It all sounded great and made the trip a bit more interesting. We were also told that Jaguar North America would send a couple of the Marketing team along as well to assess the cars and add their views on our market changes. Well, the more the merrier I say!

Having sorted out the logistics, car shipments and flights we ended up back in Mexico once again and with the benefit of being here before we ploughed into the team brief that night and readied ourselves for some altitude testing in the city the next day with our test cars that included not only the new S-Type but also a BMW and Mercedes as a benchmark. The testing was quite an easy route around the city in traffic to see how the car performed followed by some heat soak tests to let the car get hot and then see how they re-start but the main part of the testing was to repeat the 0-60 tests and see how they got on in the thin air. The changes worked well and the car felt really good which is always a blessing when you have a car magazine with you! The previous trip was really useful and we learnt a lot from it that put in the correct actions for a specification change for that market but what I didn't realise at the time was that the lessons learnt on this test would also come in useful some twelve years later for a slightly covert operation into Denver Colorado (see chapter 19).

Having subjected all the cars to some heavy city traffic driving and then did the 0-60mph tests at a closed circuit our Mexican hosts invited us to lunch as guests of what must have been the Mayor. We were instructed to follow the armoured security truck and it set off to a village on the outskirts of the City through narrow dusty streets. It was down one of these streets that the black security truck stopped and I started thinking about being trapped and ambushed. All the fears highlighted by Ford security were now flashing up in my overactive imagination but my early concerns were diffused when two huge gates opened and we drove in. On the other side of the gates it was just paradise. Lush green lawns stretched out from the archway we had just driven through bordered by thousands of flowers and plants. After being in the city the palatial garden was quite literally a breath of fresh air.

On the lawns were peacocks stretching their legs and fanning out their amazing tail feathers. This was a proper oasis in every way. We were led through the gardens to a seating area under the beams of a massive pergola that could seat twenty people easily. We were seated and the starters were delivered to the table. Now I have to say I adore guacamole so when one of the starters arrived, that was just a round shaped taco full of guacamole and topped with tomatoes, I jumped over the table to ensure it was mine. I proceeded to devour this fantastic starter and in about 5 mins it was all gone. Our host looked over to me and asked if it was good? 'Splendid just splendid' I replied. Oh good because your guacamole starter was in fact just a taco dip for the table!! I was so embarrassed and unfortunately the team latched onto this and never let me forget it. It was a great team meal in an amazing setting that set us up well for the next stage of the trip. I did however make a schoolboy error in eating some salad that had been washed in local water and that error would come back to haunt me in in the next day or two.

We left the compound in our fleet of cars and headed out to the Acapulco highway. It's a fabulous stretch of toll road that only the rich use to get from Mexico City to the beach and its practically empty again so we are able to press on a bit in all the cars and enjoy some high-speed assessments. Richard Bremner is enjoying the drive and had already suggested we do a photo shoot on route to Acapulco. Our Mexican support team suggest a place called Taxco famed for its silver jewellery production as well as some Spanish colonial architecture so with sat navs set we head for the hills. It's a fabulous day with the sun beating down and having left Mexico City we are now seeing the real Mexico.

We are soon heading west to Taxco and swap the fine tarmac road for smaller winding roads that lead into the town and as we get near to the centre we notice that the whole town must be out for a celebration. As the cars roll in there is a lot of finger pointing and I assume out here they may think we are either secret service or drugs barons. We park up the cars and are soon surrounded by hundreds of children. Our security man tells us to find some gifts to give the kids and luckily we have a few bags of sweets in the car. My support team from Mexico are already chatting away with the locals explaining why we are here whilst I start passing out the sweets. The kids are loving it, but our fashion come car photographer is loving it even more snapping away and getting some great pics for the magazine. One small girl approaches me with a basket and is trying to get me to engage with a small cigar shaped tube made from dry grass. The team give me the nod to go with it and she gets me to put a finger in each end. As I then try to pull them out it gets tighter and I'm stuck and the only way to get out is to buy the offending article for a dollar which Steve has at hand. As a treat I let her sit in the car so that her family can take pictures. Whilst we have been entertaining the locals the guys have located a box of baseball caps that we put in the car to give away and we are soon swamped again as they all want a Jaguar hat. It's just fantastic to have moments like this. An unstaged venue with spontaneous reactions from the townsfolk who are loving the two Jaguars. You couldn't buy moments like this and when we checked the time and decided we had to leave I was quite sad. I could have stayed all night and enjoyed the celebrations with them, but we had to get to Acapulco before nightfall.

Back on the highway we pressed on at pace again and soon reached our destination for the evening. We parked up, checked in and hit the ocean again for a quick and refreshing dip. The quick dip saved us having a shower and enabled us to all get ready quickly for a night out in Acapulco's swankiest restaurant overlooking the bay where they do cliff diving. It was a perfect end to a perfect day.

In the morning we had two tasks. The PR machine needed to get some glamourous photos of the S-Type and I needed to do the 0-60mph tests again because we were now down at sea level. We opted for photo's first and took the cars down to a lovely little fishing harbour and managed to persuade the locals to let us put the car on the boardwalk where we even managed to talk a local fisherman into posing for us by the car whilst he tended to his fishing nets. The photo op was just perfect and made it into the magazine but it's a shame we didn't get a picture of what happened next. The ground shuddered for a few seconds and the locals told us it was an earthquake. They switched on the news and after a while they confirmed that it was just off the coast and we should move the cars as quickly as we can. We rush around like headless chickens and eventually get the cars off the boardwalk and up to some higher ground just as a miniature tsunami sweeps into the bay. There was enough of a swell to have taken out the cars so we were quite lucky. Being lucky is again repeated on a later trip to Australia so it must follow me around!!

With PR satisfied I lead the team out to a nice flat piece of road and get the cars set up for some standing start tests. I let Richard do the driving so that he can see how much the cars change at sea level and it makes a massive difference knocking a few seconds off the stopwatch. The Jags have data loggers fitted so I download the data to take back to Powertrain at Whitley so they can validate it and sign off the modifications they have made to the calibrations.

I enjoy doing this work so much as it gives you an immense amount of satisfaction in completing the tasks you have been given by the project team, providing good data, helping to sign off the car and lead a mixed team of Engineers, US Sales, PR and a Journalist on a trip they all loved and will never forget. When I submitted my final report a while later I was very proud of what we had all done.

The drive back to Mexico City was a breeze and I swapped into an XK Convertible with the photographer to get some action shots of Richard driving the S-Type. Being the driver doing action shots is something I end up doing a lot and I get quite a kick from it. You need to be smooth and listen to the photographer a lot. The key to this shoot is to let Richard stay at a fixed speed and we move around him to get all the angles. Its great fun, especially at speed!

We enter the City and go to refuel at a petrol station that has armed guards and all too suddenly the real Mexico has gone, and the heaving metropolis sucks us back to reality.

With the cars put to bed the team decide to have a leaving meal and let our hair down a bit as PR are paying! It's another great night of banter and bonding that is all too much for Jonathon our photographer who falls to sleep like a baby after a hearty feed of warm milk. It's an opportunity that we can't miss so at the end of the meal we get all the waitresses in the restaurant to sit around him and get one of the waiters to wake him up demanding the bill. The look of panic on his face was so funny as he came to terms with the fact that all the team had gone and he hadn't got his wallet. We were watching from the other side of the building but returned to pay the bill and save his bacon!

The next morning, we were all due to go out to the pyramids but that washed salad I had the previous day had come back to haunt me. I passed on the trip out in favour of spending the entire day strapped to the toilet trying to control my ever-exploding backside. There was no sympathy and they all left to have a smashing day out like a group of tourists do and when they finally returned, I had lost about five pounds in weight and was looking as rough as a badgers arse. I let them have their banter and laughs at my expense but at least I had my Mexican revenge on the ground. On the flight back to Phoenix the next day poor old Richard had to ask the stewardess if he could swap his seat for a permanent position in the toilet. Good job it was a short flight I say.

Adios to Mexico and onto the next adventure.

# 8. Spinning Around In Sweden

So, the long and short of it is that I have never been out to Sweden to do any engineering testing at all but as always, I was lucky enough to be volunteered for two trips that involved filming.

Sweden is a beautiful country and in the winter months we enjoy about twenty weeks of Cold Climate Testing out there. On my first trip I was working on S-Type and PR decided they would like some video footage of the cars being assessed and how the traction control teams were busy getting the calibrations sorted out for the car.

The idea was that myself and Gary Aldous from Chassis Engineering would assess the vehicles to make sure they worked on the snow and ice correctly and give some non-biased feedback to the teams. After this we would be filmed out on the ice lake doing accelerations, braking and cornering with traction control off and then switched on. With the systems switched off the cars were all over the place spinning out and being very difficult to drive on the ice. With the systems switched on they transformed and behaved so well giving the driver a lot of confidence. It was during the filming though that we really upset the traction control teams as the film crew wanted us to do a transition from poor calibration to good calibration, but we didn't have the hardware to do this out on the lake. Gary and I decided to use a laptop and a lead and make believe that it was hooked up to the car. I then drove around the ice lake making sure it was sliding around and behaving badly. When we stopped Gary would ask me for my opinion and then make it look like he was changing the calibration before asking me to try it again. I went back out and drove a perfect lap and back on film I commented on how good his changes were and it was a huge improvement. Whist this in effect is what the guys do it does take a lot longer which would have been rubbish for the film, so we went off piste and faked it a bit.

The film crew were happy but the traction boys were livid that we made it look far too easy! I seem to remember we had to buy them a lot of beer that night and at Swedish prices they made the most of it.

It was also another trip where you get to bond with the team and establish relationships that carry over into other projects. On this trip I got to find out so much about traction control systems from Andy Gosling and to this day he is always the person I go to for skunk work calibrations as I did with Project Bonneville, XKRS-GT and Pikes Peak.

The bond with Gary though was taken to the next level when we decided that after a hard days driving we needed to recover Swedish style with a relaxing soak in the sauna. In our swimming trunks we entered the steamy vault and picked a space on the pine bench and as the steam cleared a bit, we realised we were not on our own. There were two ladies present in which was obviously a mixed sauna but as they turned to greet us it became apparent that being naked was the norm. They gave us the look, I gave Gary the look, he raised an eyebrow and that was that....when in Sweden do as the Swedes do! I think we sat there silent for twenty minutes trying to focus on anything else other than being in the situation we were in. These two were just like ones from Abba, not the ones with beards either. They engaged in a little conversation, but it was eyes forward at all times and just pretend they are not naked at all. Thankfully two gents entered and before the towels were removed we were out of there.

On a short trip like this Gary and I have a lot of fun with the team but on longer trips they struggle a bit because there is not a lot to do in the cold winter months and when you go out to a bar you are surrounded by other teams from the motor industry so you have to be so careful with your conversations and not let anything slip about what you are testing.

My second trip into the snowy world of Sweden was for another bit of film making with the brilliant XF. Our Sales and Marketing Team were keen to get a good short film to support the motor show launch of the new car and having dabbled a bit before with S-Type I was put forward as the driver and support engineer and promptly despatched out to the small town of Kiruna for the week.

Kiruna is in the far north of Swedish Lapland that was famous for its iron ore mine. It has a population of around 19,000 and was once listed as the largest city in the world by area although most of its space was non-urban.
It's also located about 90 miles north of the Arctic Circle and has snow from about September through to mid-May which makes it a great place to conduct our winter testing. Those unlucky enough to have done any late testing from about the 11th December to 1st January would have to cope with no sunshine at all and the average temps for the winter are about -30°C.

The base in Kiruna was run by a good mate who I have known for a few years called Phil Talboys who bases himself there for the winter months to control all the Cold Climate durability testing before heading back to the Nurburgring for the rest of the year. It's one of the top jobs and he loves it most of the time!

On meeting the full team for this trip I realised it must be a big production as we had a producer, a couple of cameramen, sound engineers, a co-ordinator and a film technician. They were all good sports and I became friends with the senior cameraman Casper Leaver. Casper had been doing this for a few years and his claim to fame was that he was part of the Top Gear team filming all over the world. We hit it off straight away which made the filming a lot easier to be honest because every time the producer told me what he wanted Casper would translate it into simple engineering speak like "Just drive really fast, slide the car sideways, create some snow flumes and smile".
Just how cool is that? Being asked to have fun and being paid to do it.

During one of the days I had to do a lot of drifting in a circle and it was a gorgeous sunny day. I parked up at one point and

Casper and I decided to walk around the ice a bit and to my amazement and horror the circle that I had been using was getting warmer with my tyres spinning on it and also the sun was melting the top a bit. You could see flowing water underneath the ice. We had to change our location after that, but it did make you realise that this is actually quite dangerous whilst also being just brilliant!

I did most of my training at MIRA on the wet grip circles which are made from Bridport and Pebble with water sprayed onto the surface making the slate type materials feel like ice. When you go out onto the circles you accelerate sharply and it kicks out the back end quickly causing you to spin if you don't catch it. I spent hours and hours on this and the key to smooth driving on the wet surface was smooth throttle inputs and smooth steering inputs.
Keeping the car drifting sideways, once the steering was set up, was through the accelerator pedal and getting the right balance. I enjoyed it so much that I ended up doing some driver training for some of the project teams so that they could feel what it was like when a heavy saloon starts sliding on a slippery surface.

Being filmed doing something you enjoy, in a vehicle you have helped develop for one of the most iconic brands in the world, is just so special. I always felt honoured to be in that position and I always gave it 100% and I think the film crews enjoyed that as well. When I needed to I remained focused on the task and made sure they got exactly what they wanted but away from filming I always made sure the team had a laugh and enjoyed it as much as I did.

I learnt early on that it's always best that whenever you are asked to do something you say 'Yes, of course I can do that for you" as it creates a better working environment. So many times in the business I have heard people say "No I can't do that" or found a hundred reasons why they can't do something, and it creates a bad atmosphere as well as being counterproductive. If you say yes but then fail for other reasons people are generally more accommodating. So, when I was asked to drive as fast as possible on a snow road to create some rear end snow flumes it's a simple and very accommodating answer, "Well yes of course I can, but how fast do you think it needs to be?"

The answer was well into three figures and although my initial thoughts aired on the side of caution the acknowledgement that the top speed would be completed on studded tyres filled me with far more confidence.

Studded tyres give you huge amounts of traction and it only took a couple of runs to get well into three figures.
Having established the camera positions a covering of snow was added to the track and I was again released from the makeshift start line. I hit the snow mounds at about 120mph and the rooster tails from the rear end turbulence were so impressive. The team loved it and used it in the final edit.

With a successful video completed and a very happy producer it's only right that we end the trip with a team drink and I suggest that we go to the ice hotel at Jukkasjarvi.
We arrive at this amazing place and start taking lots of pictures having a few laughs and generally unwinding when Casper decides to put his tongue on an ice sculpture and gets stuck. He starts making all sorts of strange noises and it's only right that we leave him there for a while until we all have a lot of pictures of him. Poor chap had to wait a while before we found some warm water to get his tongue off.

The actual hotel is just fantastic and it takes them a few weeks to put it together. They harvest the ice in March and April and put it into cold storage before sculpting it in November. At the end of the season it just melts and goes back into the river.
We have a brief tour around and even take a look inside an ice bedroom that looked so cold. You have to wear special gear for the night and wrap yourself up in reindeer pelts.
Our destination though is the bar which is also a gorgeous sculpture and we relax in ice chairs and sofa's sitting on top of more reindeer pelts. There is a great ambiance in the place and as I'm feeling good about the trip I suggest that I get the first round in with Vodka for all eight of us but that good feeling flew away when I got the bill of 80 quid!

The vodka was served in carved ice glasses that believe it or not you could actually keep and take home with you in specially prepared boxes so they did not melt. When the barman asked me if I would like to pay for this service I gave him the most serious of raised eyebrows I could....he got the message!

The final video from the trip was used as a backdrop to the launch of the car at the UK Motor Show and when I watched it play I was so proud to have done my bit with the team.

I have to say that when I talk to people about this experience it's the one that makes them sit back in horror and question the stupidity of it all. The weight of the cars, the speeds, the heat and wear of the ice and the fact that if it cracks you are going under. I can quite honestly say that I understand where they are coming from but the vast majority of them are not engineers and that's where we are different.

As an engineer the most important question is "what are the requirements?"

Non-engineers would look at the driving on ice task and try to figure out how they could get on and off the ice without causing it to crack and then perish in the icy cold water.

My engineering view on the same question would be "how fast do you want me to go? How big do you want the drift to be and can you send me a copy of the video?"

So.....to my girls. I had the most amazing time doing this on a frozen lake that had a minimum thickness of 8" and was checked every day by the road administration and ice test team. We followed very strict guidelines paying attention to the surfaces and pressure waves that a car creates on the ice. We never parked cars together and never drove them alongside each other.

We always expected and planned for the worst and made sure the window was down at all times so that you could jump out.

As engineers we have very pragmatic views on what we do, we look at all the relevant data and then make our own risk assessments. With all those boxes ticked we then have an absolute hoot!!!!!! However, if I ever catch you skating on frozen canals or lakes in the UK there will be serious trouble!

# 9. Going South to Atlanta

Well things were going well at Whitley and I was really enjoying heading up the competitor vehicles department where I had access to all the cool cars and was invited to all the project ride & drives as a neutral and maybe the voice of the customer. As a company we were well into the relationship with Ford and had just announced that Jaguar and Ford would be sharing a platform with the Lincoln Project Team in Dearborn to do a New Lincoln and a Jaguar S-Type.

The competitor drives in the UK were focused against the latest BMW's & Mercedes, but the combined project teams decided it would be a great idea to do a shared drive event in the USA.

Vyrn Evans who was heading up the S-Type Vehicle Engineering Team asked if I would help co-ordinate this on their behalf and have an input with the cars we would compare and then work with the Lincoln team to work out drive routes and event logistics.
Part of my role in the UK was to also run the Vehicle Assessment Group that drove the cars and assessed how good they were. The VAG team were made up of people from around the business such as Manufacturing, Purchasing, Quality, Engineering, Finance, Powertrain and Programme Office who would rate the attributes of the car versus the competition to see if we were better or worse. For this drive the project team also wanted a cross section of the VAG team to attend the drive.

We had a few debates about who was on the drive that took days to agree before we all sat down to decide the location. Atlanta Georgia was selected. I have no idea why and I can't remember for the life of me, but it must have been a good reason I'm sure!!

All of the cars were being supported through Ford Dearborn, which included a new one for us the latest Nissan Infiniti, and when they had arrived both teams would fly out to Atlanta to join them. As I was going to help my counterparts in the USA to set up the event I flew out a couple of days earlier with my boss Peter Fawke and the Programme Manager Peter Matkin. Both of these guys were also very close mates, so we were all looking forward to the trip.

The rest of the team would include the S-Type Senior Manager Vyrn Evans, Programme Director Nick Barter, and VAG Team members Alex Chamberlain, Dave Rooney and Mike Langdon from the UK and then all our opposite numbers from Dearborn in Michigan. Interestingly there was a late stand in for our Powertrain Senior Manager Ron Lee who could not attend so he was sending out a V6 engine calibration engineer by the name of Richard Chant.

I knew Richard from previous testing, and he was an excellent calibration man and good bike rider but I didn't know what he was like with vehicle assessments but either way he would be a good addition to the team for his knowledge on V6 engines.

Well the three amigos arrived in Atlanta and the first thing Matkin wanted to do was some clothes shopping in the terminal mall to get a couple of shirts. He was so fashion conscious it was untrue. I once totalled up the value of his Hugo Boss suit, Saville Row shirts, Breitling Watch and a pair of Grenson Brogue shoes and it came out way more than the cost of my car at the time. He did always look the part and his choice of watches did rub off on me in the end.

Fawkey aspired to be like Matkin but not in his price range but always had an excellent choice of suits from some up-market shop in Birmingham.

The first port of call on this shopping trip would also be my last. We entered the world that is Versace and wandered around whilst Matkin decided what he wanted. Fawkey and I were just window shopping really and stuck together. We decided to try and find something that we could afford and maybe buy just to say we had been there, so we headed over to the belts rack. I started looking at some nice belts and selected a fine-looking heritage tan leather belt but on close look at the price tag it came to nearly five hundred pounds. I turned to Fawkey and in what must have been a loud and shocked voice I screamed "Check this out mate, five hundred quid for a flipping belt and all it does is hold your trousers up."

I found it quite embarrassing being escorted out by security but not half as embarrassing as it was for Matkin being associated with us. Fawkey and I could see the funny side but Matkin ended up wittering on about it all week!!
A great start to the trip in my books.

We were taken to our really nice hotel and checked in quickly as always and with nothing to do that night the plan was to meet in the bar in fifteen minutes. I arrived at the bar and the first topic of conversation was the unlimited amount of TV channels and in particular the range of adult only channels.
Even in the fifteen minutes Fawkey had scanned through a few and mentioned that each one had five minutes free. I commented that if the drive event was no good for thrills then the TV would be a good back up.

The following day he came down to breakfast looking a bit sheepish and I asked him if he had watched all the free movie trailers. He advised us that he did watch a few but he allowed one to go past the five mins by a couple of seconds and the TV system had now released the entire video to his room number. His sheepish look was because he had just found out that the room bill details the films you have been watching and how would he explain that when our bills were signed off. Matkin and I were in tears for ages as Fawkey ranted about it all night.

Amongst all the fun and banter we also took delivery of the cars, set up a suitable road route to take us over various road surfaces, inclines, descents and a combination of freeways and minor roads and also managed to set up route drive books for all the guests. Our colleagues from Dearbon joined us the next day for a dry run of the event before the seniors and execs arrive the following day to assess the cars.

When I woke up for the early start there was a very strange phenomenon outside that Brits would call a light splattering of

113

snow. Nothing too serious, just a dusting. I cleared the windscreen and took one of the cars to get it fuelled up. On route I listened to the local radio news and it appeared that there was total chaos of the highest order. The traffic presenter was raging on about how many crashes there had been on the freeway and the news reporter was reading out a list of all the schools that had been cancelled. It was absurd and I sat shaking my head before I lost it with the radio and shouted back that it was JUST A DUSTING and you all need to stop ranting on and get to work or take the kids to school in your army of SUV's.

I drove back to the hotel inducing the odd tail slide for maximum effect as you just had too. It was great fun too.

The next casualty of the 'slight dusting' was the airport and having announced this the place went into travel update overload. It was hilarious and I engaged our receptionist on the very subject, but she informed me that it was a serious problem because the vast majority of locals had never seen snow before let alone drive in it. Well I guess that's a good contributor to the pandemonium down the street!

Our colleagues from Dearborn also joined in the madness and were keen to cancel the drive but we were having none of it. I ended up agreeing to do a drive of the full route with one of the American team to make a judgement call. By the time we finished the route the dusting had gone and the sun was out just like any normal British day.

Our seniors arrived and the drive started for real and we had arranged driver pairings for the trip making sure the right people were together. The last pairing we did was to put Richard Chant in with Peter Fawke as Richard seemed worried about his role in this as he had never done it before, and Pete was the right man to make him feel relaxed and guide him through the assessment rating sheets.

We all set off on the route and arranged to have a number of driver stops to swap over drivers and then to swap cars. All was going swimmingly well until Fawkey and Richard didn't turn up at the next stop. We waited for a while but then got a call to say there had been an accident. We went back down the route and found the very new Nissan Infiniti upside down in a ditch with both guys sitting at the side of the road. Richard was driving and he came up to an incline to the brow of a hill and the road appeared to go left but just after the brow of the hill it turned sharp right and then left.

As Richard hit the brow he was too fast to get around the right hander and rolled the car over. They were both okay but Richard's pride was hurting a bit and things were about to take a turn for the worse when the State Trooper arrived. We talked him through our activities and the situation to which he apologised to us and then told us he would have to arrest Richard as part of the process. Whilst we were all outside the car it was indeed very serious but as soon as the doors closed on the trooper's car we burst into fits of giggles at the expense of poor old Richard. He was released after giving a statement and by the time he got back to the hotel we had recovered the Nissan and arranged for a replacement vehicle.

The actual drive went very well and both teams agreed on the benchmark targets for our new products and as always it was good for team bonding too, although on some things we just had to disagree with one subject standing out more than others. As part of the drive team building, we decided that the USA team would host the Brits and do a meal and select a good comedy show to watch on video and then the Brits would return the favour on the second night. Our hosts kicked off in typical style with BBQ food, burgers, hot dogs and wings that was followed up with a show called Seinfeld. The Brits sat for an hour in silence as the USA Team howled with laughter. We just didn't get it and it was not funny. Obvious humour must just work for them I guess but we need something much better. To start the proceedings, we managed to get fish and chips for the meal and for the hour of entertainment we even managed to get hold of a couple of episodes of "The Young Ones". To our amazement the USA team did actually laugh in horror at what we can get away with on TV whilst we fell about laughing for both episodes. Our conclusion from this was that although we are separated by about 3500 miles and we are so closely related in our blood lines when it comes to humour, we are light years apart.

As trips and benchmark drives go it was a fantastic success but the best bit for me came on the last night when we were relaxing in the hotel spa and I was joined in the Jacuzzi by the programme Manager Vyrn Evans. Now don't get me wrong, Vyrn was a top bloke it was not his presence in the Jacuzzi that was the best bit. It was the conversation we were about to have!

We had a great chat about the event going through all the good and bad points and then we started to discuss my career within Jaguar and how I ended up in Competitor Vehicles. After about forty-five minutes of interrogation Vyrn asked me if I would like to consider a move into the S-Type Vehicle Office to look after the development of the new cars working for Peter Fawke. I considered it for a couple of minutes before accepting the offer and never looked back. I do often wonder how many people in the world have ever had a job interview in a Jacuzzi and I would imagine it's not a lot.

As for the other two amigos, well, Peter Matkin also got some news from the Chief Programme Engineer Nick Barter as we were boarding the plane to come home. He was told it was an excellent trip and he hoped that Peter had enjoyed it because he would be stepping up in his career and therefore there would be no time for flying around the world.
He did step up in the world and was very good at it. Unfortunately, he decided a new role was required and went off to China to head up the engineering development for a new car company where he still is today.

Poor old Pete Fawke was diagnosed with stomach cancer on our return and my role into the Vehicle Office was brought forward as I stepped in to replace Pete whilst he was having chemotherapy. I was shocked by the news and doubted I could step into his boots and keep the vehicle development running as well as he was doing it, but I did. During his treatment I went to see him at the hospital at his darkest and lowest point and I thought I would never see him again, but fortunately he was a fighter and pulled through to beat this awful disease. He returned to work and over the coming years was given the all clear and I worked for him on S-Type for five brilliant years. During that period I learnt so much from him that has carried me through to where I am today.

Pete like so many others in our era took early retirement in March 2019 and like the business, I will miss him.

As for Vyrn, well he took retirement at the end of the S-Type programme having had an amazing career within Land Rover and Jaguar. He looked after me well over the years and gave me the chance to move onto the XF Programme.

We had some amazing times, some epic drives (often in Wales) and had a lot of fun developing a great car.

Down under in Australia is the proof of that!

# Part 3 – Events of Biblical Proportions

Over my thirty years of testing there have been a number of events that have just been on a different level that I felt needed their own section of the book.

These are the things that just don't happen every day which makes them very memorable.

The events include crashing a car, meeting Kings, pushing the limits, being on film and TV, spending time with one of the most famous men in America, being a chauffeur to the most successful football manger ever, breaking speed records, designing my own cars and reaching the peak of a mountain.

The next ten chapters might never have happened if it wasn't for the understanding of a couple of senior managers, a letter to the chairman and the successful launch of the XJ Saloon.

If it had gone in another direction then chapter 10 could so easily have been 'My Last Day'!

# 10. It Always Happens on the Last Day

Being in a test and development department means that you are testing cars to the limits often trying to make things break or proving that they are robust and signed off for customer usage. During the testing it's inevitable that there will be mishaps, accidents and the odd crash for various reasons and during the early years of my time in vehicle proving I was made aware of several them.

The first one I was aware of was by a chap they referred to as Dangerous Davis or Dange. He was a good mate of Steeley and I was introduced to him whilst testing the suspension on XJS and Pete was working in chassis development. Dange had been getting a car ready at Browns Lane and was having the brakes bled on the ramp.
When completed he jumped in the car, he reversed it off the ramp before the technician could tell him that the pedal needed to be pumped a few times first. He flew backwards for a few feet and when he went for the brakes the pedal went to the floor and the car pounded into the metal shutter doors. The foreman shouted, "your flipping dangerous you are Davis" and that stuck for the rest of his career at Jaguar.

Another crash I was made aware of involved one of the Vehicle Proving Test engineers Roy Toll who had a massive high-speed crash at Nardo in Italy. As a result of his crash he was mostly covered in plaster and in traction for a few weeks. When our senior manager visited him in the Italian hospital, he asked Roy if there was anything he could do then please ask. Roy called him closer and said there was something very delicate that he needed some help with to which his manager said, "fine just ask." Roy told him that he hadn't had a pee for hours and could he hold the cardboard container for him. The container was put into position and Roy let go in a big way and couldn't stop. He said that considering the situation he was in it was the funniest moment watching his senior manager panic as the container flooded all over his hands and shoes.

There were many others and a few people in the office managed to get the dreaded steering wheel trophy. The trophy was a plastic steering wheel that had your name printed on it to mark your achievement of damaging a very expensive test vehicle. As the wheel was being presented in the office on one of the days, I remember thinking just how lucky I was to be in this job and still have a clean sheet but that was all about to change.

At the end of the new XJ Supercharged programme I was asked by a Jaguar legend, Peter Leake, if I would support doing the vehicle launch at MIRA. My role was to arrange a ride and drive event so that dealers could take the new XJ out on the road and see how good it was as well as doing some high speed laps of the Dunlop circuit taking round two or three at a time so they could feel just what it was like at the limits and experience just how good the car was.

The event was over six weeks, and all was going well as I drove into MIRA on the last day. As I approached the gatehouse my old mate Dangerous Davis was just leaving, and we exchanged some comments about how well the drive was going and the fact that it was the last day. He said "well Moorey, be careful because if anything is going to happen it will be when you relax too much on the final day."

On the event I think we managed to get five hundred people through the cars, and I must have done about three hundred and fifty laps of the Dunlop circuit going through many sets of tyres in the process. During the circuit drive I would collect three people at a time and then take them up to the track and do a brisk first lap and then a flying lap pushing the car hard on the corners to show how well the tyres gripped and just how balanced the car was. So, we arrived at what was not just the last day but also the last group for the circuit drive and I collected three passengers from Jaguar Coventry. I will never forget the manager for the group who was called Noel Gaston.

He buckled himself in and we did all the introductions on the way up to the track. It turned out that they had heard a lot about the track drives and were so excited to have the chance for a couple of laps. Noel also told me he shared a love of motorbike racing and was a bit of a speed freak and don't be afraid to hold back.

With his comments noted and on the basis it was the last day I went out for the first lap feeling pretty good about life.

We did a very quick lap and had the car drifting through the long bends to the delight of the passengers as we then sped past the control tower for lap two. On this lap I made them think we were heading off down the straight but cut through the infield using the 'S' curve which caught them by surprise as to just how nimble the big XJ was.

With a nice slide out of the final bend we finished the lap at the control tower and I parked up to open the gates and let us out.

As I got out Noel joined me and was high on the adrenaline rush and could not stop enthusing about how good the car was and what an epic experience he and his team had just been part of. He looked at his watch and told me they still had ten minutes and could they do another lap. I had a look at the tyres and said that we should probably leave it as they had seen better days, but it didn't wash and I ended up with the three of them pleading for an extra lap.

At this point I made a wrong decision! I should have said sorry and driven them back to the workshop but in the euphoria of the moment and with it being the last day and the last set of tyres I thought one last lap will be fine.

We set off from the control tower and slid the car around the long-left hander and I decided that we would go straight down this time and into the second long left hander before heading up to the tower. As I lined it up for the turn I let the car settle before giving it some power to start the slide and as I had done for the previous three hundred and fifty laps I let it drift to the edge of the track but this time it went maybe a foot further and onto the grass and at that point it was game over. The tyres were no good at all and the well balanced XJ Supercharger became a V8 helicopter. The backend swapped with the front and we span across the grass and hit the tyre wall before being thrown

121

into the air landing back on our wheels and coming to a shuddering stop back on the track. I immediately switched my attention to the three passengers to make sure everyone was ok and then got them all out of the car as quickly as I could.

With the passengers checked out and returned to base for the coach trip back to Whitley I returned to the track to take a look at the car with Peter Leake who was trying to keep me calm and assure me that everything would be ok but on seeing it sitting there with every panel dented, flat tyres, broken suspension and smashed lights I nearly passed out. Peter was awesome though and took me out of the loop straight away for a coffee and he continued to re-assure me that accidents happen.

We called Browns lane a while later to see what the car was destined to be used for next and that did not make things better at all. The car was due to be checked reworked and then polished to go on show at a dealership in London the following week. My heart sank and I have never felt so low in all my life. I had let the company down, the event down and let myself down as well as scaring the living daylights out of three dealers.

I left the track later that Friday and went home to tell my parents the bad news. They were also supportive, but it didn't help either and I went to bed. I don't think I slept a wink that night or the Saturday, or even the Sunday come to think of it. I tossed and turned each night replaying the moment repeatedly. How Formula 1 drivers have a massive crash and then just race again is beyond me.

I arrived at Whitley on Monday morning expecting the worst and I took a deep breath before entering the office. I was met by many of my workmates and as always engineering banter took over and there was a lot of ribbing and mickey taking as I made my way to my desk where the plastic steering wheel trophy was standing tall. My name was added to the base next to the title of 'Crash with Panache'.

Having dished it out a few times I just had to man up and take it all in even though in my mind I just wanted to throw the flipping thing through the window.

I ran through the story a few times during the morning before I was summoned to the Senior Managers office for a debrief and the reading of my last rights with Andy Wheeler. I entered the office to see my boss Dave Hopkin with Andy and the Sales & Marketing Exec Stephen Perrin.

As I stood there shaking Andy asked me to explain in my own words the events of the Friday drive and I went through them in every detail. He asked what I had learnt from the experience and what should I have done and be expected to do in the future. I had obviously gone through this a million times over the weekend and I responded quickly. Having listened to me for about ten minutes he nodded and said that Stephen would like to say a few words. Stephen stood up with a sheet of paper in his hands and said that he would like to read something out.

"I have here a letter from Noel Gaston head of Jaguar Coventry written to myself and copied to the chairman. Noel explains that he would like to take full responsibility for the accident having put pressure on you to return to the track. He said you had advised them all that you should leave the track, but they put the pressure onto you for an additional circuit. He continued to say that up until the crash you had demonstrated the vehicle at the highest level and apologised for making you continue."

Stephen put the letter away and then did a lovely speech about the success of the event and how many dealers had contacted him to say they had a brilliant day and how stunning the car was. He then said that all the people he had spoken to had commented on how good the drive event was and how much they loved the track drive experience.

On that basis they agreed that there was nothing else to say on the matter and I was told I could go. Many years later I still remember this every time I do an event and I learnt a lot from the experience.

The plastic steering wheel trophy only lasted a couple of days on my desk thanks to my mate Duncan Mackenzie who managed to jump out of a car thinking he had put it into park but left it in reverse. When he returned the car had gone and he wondered why a visiting test team where standing around a car near the fuel pumps. At this point he realised that it was his car that had embedded itself into a parked car having gone backwards from where he left it. It was with great pleasure that the trophy was handed over to a new owner.

Having finished this chapter there are a number of other little moments that are worth a mention in dispatches.

Terry Igoe for preventing a field of cows from running out into the traffic having left the road and put an XJ through the farmers' fence!

Steve Peat for turning a Jaguar car into a boat getting it stuck under a flooded bridge at MIRA

Mark Evans for closing the M40 after spinning out my only XKRS-GT into the armco barrier.

An unnamed Exec who put the on-loan Merc SL into a ditch and left me to sort it all out with the press and police.

Dave Gaylard for ejecting a water filled dummy through the windscreen on a pothole brake test.

Kev Gaskins for putting an S-Type through the workshop double doors and into the main corridor.
(Pressing the accelerator and not the brake pedal was a huge mistake).

Graham Wilkins for doing an impossible photo shoot on a wet beach and getting an XF stuck.

Richard Chant for an epic roll over in the Nissan Infiniti.

The Bilstein damper engineer who ran out of talent on the A444 and parked the XJS sideways into the garden of the Red Gate Pub.

My summary for this chapter is a quote from Theodore Roosevelt "The only man who never makes a mistake is the man who never does anything."   A wise fellow!

# 11. Unexpected VIP

As far as unexpected guests are concerned this one came in at number 1.

We were just at the end of the XK development programme and as part of the sign off we had Sir Jackie Stewart in attendance at MIRA and quite a few Execs to assess and sign off the car.
My role was to ensure the cars were fit for purpose, book the facilities and run the event and to be honest it was a bread and butter day…..right up to the point Sir Jackie announced he had a VIP guest joining us after lunch. He handed me a piece of paper with a telephone number on it and told me to call them and sort out getting him into MIRA.

As the team assembled for the technical brief I go back to the office and ring the number with no idea who this VIP is.
I finally get through and I'm asked who is calling….I give him my details and tell him I'm from Jaguar Cars……"Oh Yes, we have been expecting your call. I'm airborne at 11am but can you get me clearance for arrival"

"Erm…….did you say airborne?"

"Roger that. Do you have co-ordinates and clearance for MIRA approach?"

Trying to sound like this sort of thing happens every day I confirm that I will get the co-ordinates and clearance and call him back.
I trot over to the control tower and luckily they have a landing spot in the middle of the test track and a temporary 'H' symbol which they will set up for me. Armed with appropriate grid ref I give the unknown helicopter man a ring back.

He is pleased with the info and confirms his ETA.

Before he goes I ask who the VIP is….there is a silence……followed by "I'm sorry but I'm not authorised to give you that information"…….Wow…..this is big!!

I pass on the information to the drive team and I'm then told that my new priority is to make sure we have two chauffeur cars at the control tower for the arrival and support any requests thereafter.

My bread and butter day is now a whole lot more interesting. Without any delay I'm in the back office with all my old mates passing on the details so that they are aware of what's going on.
We spend some time speculating on the identity of the VIP in the hope that it's some major superstar, which is highly likely being a friend of Sir Jackie!

At 11am there is a call from the office…… "Phone Call for Mr Moore Urgent"

I take the call…. "Hi David, just want to confirm that we are now airborne and in transit"

Crikey……how cool is that……the helicopter captain is calling me from his aircraft……I love this job!

Right…. chauffeur cars…..like I have cars just sitting waiting for this sort of thing……I wish!

I make a few calls and call in some favours and in no time at all I have two nice XJ Saloons in pride of place at the control tower.

The control tower at MIRA is the original tower from when the test track was an airbase during World War Two. The base was used to train Spitfire pilots and the tower had not changed since then! It's a lovely old building and it still has that historical feel and smell.

I go into the rest area and meet up with Mick the track maintenance guy who has sorted out the helicopter landing area for me in the middle of the Test Track.

"So, who is coming in then Dave?"

"I have no idea Mick……hope it's a major celeb though"

"Ok mate" Mick raises his eyebrows and makes us a brew.

As we sit and chew the cud there is a distant sound of rotor blades, so we hurry outside for our arrival.
The noise gets louder but we still can't see the helicopter….I'm expecting a nice private jet ranger or something but I'm not prepared for what's just about to pop into view!!
Just up over the trees is the biggest military helicopter I have ever seen, and it turns towards our temporary landing pad.
As it lands there is a massive swirl of dust and debris as the downwash blows everything out of sight. The jet turbines are shut down and the side doors open. I have no idea who it might be but the face at the controls looked familiar.
As I walk forward, I notice that some of the Jaguar Execs are waiting on the other side of the barrier with a number of cars ready for our additional guests so they must be quite important.
I stop just short of the helicopter as several guys looking like secret service agents exit the aircraft and they are followed by the familiar face from the cockpit.
As he approaches and shakes my hand, I suddenly realise that it's King Hussain of Jordan. I welcome him to Jaguar MIRA and then he turns to Mick and shakes his hand as well….the look on Micks face was hilarious!
I quickly hand over our guest to Sir Jackie and I'm left to chauffeur one of his sons Prince Abdullah who I get to meet again some year's later (see section 5).

As the party go for a meeting, I'm given another task from Sir Jackie "Can you get an XJ220 over here now".
My initial response was "Yeah Sure, no problem" but what I was actually thinking was "*Oh Yeah that's just so easy…we have fields of them just parked up ready to be used.*"

I call press cars and ask about the availability of a 220 at MIRA but the response of "Who are you" kind of lets me know it's not going to happen. I explain that we have a guest that would like to drive one, but that also falls on deaf ears, so it's time to pull out my ultimate top trump card......."His Royal Highness King Hussein Of Jordan would like to test drive an XJ220 please". There was a slight pause before a very nervous response of "We have a silver one if that's ok"

The car eventually arrives and the head of our dynamics team Mike Cross takes the King around the circuit at breakneck speed.
Whilst we wait, I engage with one of the security agents from the UK with a bit of idle chit chat and question him on what would happen in the event that the car might crash with the King in the passenger seat?
He looked me in the eyes and said, "Listen son, if that happens I will personally shoot the driver myself" and he looked very serious about it too!

Well, everything must have gone well as deals were done and The King then flew the Execs back to the Product Development Centre at Whitley which must have been the most expensive taxi ride ever!

It's not every day that you get to meet a king, and in your lifetime it might just happen the once. I managed to meet his son when he became King some years later, and I also got to drive the King of Sweden around the track at Millbrook. I need one more for a Royal Flush!

# 12. The Fastest Road Trip Ever

As far as events of biblical proportions go this one takes the biscuit by a country mile and due to changes in the Australian driving laws it will never be repeated or beaten.

It was also that epic that I could have written a book just about this event in the style of Bill Bryson but as it is, I'm going to shorten it a bit as another amazing chapter in my life at Jaguar.

The S-Type, like all vehicle programmes at Jaguar, comes to an end for engineering and gets handed over to manufacturing who will then take the car into production for the customer. As we were doing this transition it was coming up to Christmas and the whole of the project team were enjoying the festivities at a local golf club.

I was sitting next to Ashley Emmerson, who was the manufacturing manager, and we were both enjoying talking about the development of the S-Type over the last couple of years when he suddenly asked me what I was doing over Christmas and the New Year. I said not much really as I intended to just stay at home to eat and drink.

He looked me in the eyes and said "Oh, Good. How would you like to go to Australia and do some work for me?"

It was at this point I had to check how much he had been drinking because this was not Ash talking. It turned out he was driving and drinking only shandy, so it was a valid question. I accepted his request with the caveat that he had to run it past my boss Vyrn first but apparently it had already been discussed and he was happy if I was.

Well, the run up to Christmas was a blur as I had to sort out getting two cars to Australia, arrange for my visa and injections and touch base with the Ford Engineering Manager out in Geelong who was going to help with the work when I arrived.

The mission that I had accepted was to take two cars on a 10,000-mile road trip as the first part of a final validation of the pre-production vehicles. Due to the weather conditions in the UK it made sense to go to Australia as they were in their summer so it was far better from a testing point of view because after my 10,000 miles the cars were going to continue and complete 100,000 miles at the Geelong test facility.

What a bonkers idea and what was I thinking when I said yes? I must admit that the prospect of doing more testing is always good and going to Australia was even better but when you consider doing a thousand miles a day for ten straight days it's mad….and I love that!

So here we go again, the 26th December 1998 and I'm flying the flag again on flight BA7310 to Melbourne, and as it was a works trip that was over 11 hrs long and during the holiday period I was flying up front. For years I had been wanting to turn left when I got on a plane and here I was enjoying the good life.

There were two moments worth a mention during the flight and the first was at our stop in Singapore. I got off the plane and had a quick wander round and then sat next to a couple as we waited to board again. They were quite chatty and then there was an announcement asking if all first class and exec club passengers would like to board. They both sighed and passed comment about people having all the luck to which I replied "yeah, sorry!" I got up and boarded like a seasoned professional.

The second classic moment was when the pilot announced that we were just flying past Darwin on the left-hand side of the aircraft so I started to pack away all my stuff and put the table up and basically prepare myself for landing. The steward came up to ask what I would like for dinner and I had to ask if we had enough time for that. We certainly had enough time alright, about 6 hours to Melbourne! My god what a big country.

I was met at the airport in the early hours of the morning by the Ford Vehicle Engineering Manager Bryan Knowles and his daughter Megan. Bryan was a Brit that had moved to Australia and taken up his current role with Ford in Geelong. Within minutes of meeting him I knew this was going to be a great trip as all we talked about on the way to the hotel was the test match that had started in Melbourne. England had drawn the first test but then lost the next two but still had two to play.

Bryan dropped me off at the hotel and arranged to collect me the next day, and that's all I remember. The flight took just over 21 hours having left on Boxing Day and landed on the 28th December and I was shattered. My head hit the pillow and I was out for the count!

As promised, I was collected the next day and we made our way to the You Yangs Proving Ground about 18miles outside of Geelong to see the cars and meet some of the team. They had already been busy getting everything together for our epic journey and there wasn't much to do other than have a quick go in the cars to make sure I was happy.
The team, considering it was Christmas holidays, were really keen to crack on and get on the road asap but I told them that I still needed another good night's sleep before I could even think about driving.

Bryan had put together an engineering team with some good experience and a support team to follow us for the whole trip who seemed like they were ready for some fun too. The engineering team consisted of Bryan Knowles Product Development Manager, John Willems Powertrain Manager, John Birrell CAE Manager, Rob Hall Chassis Development, Greg Locock NVH Development and the support guys were Mark Rowlands and Russell Brame both of whom were what I would call true Ozzies who could give and take a bit of banter.
After our introduction Bryan did us a BBQ and we went through the itinerary for the trip.
The plan was to leave Geelong and head North West to Port Augusta followed by Alice Springs and then to Darwin at the Northern tip. To increase the mileage, we intended to stay at Alice Springs for a day and back track to Ayres Rock and then once in Darwin we would spend three days there to make good use of the Northern Territories lack of speed limits for some high mileage days. Our intended route was to come back down

the East coast via Brisbane, Sydney and Canberra but this would change when we got to Darwin.

As we confirmed the route Bryan advised me that as soon as we hit the Northern Territories there was no speed limit and we could make some great progress on the Stewart Highway but with two very thirsty Jaguars we would have to make sure we refuelled at every petrol station on route because they were few and far between where we were going.

Bryan also mentioned that his intention was for us to stop over at Alice Springs for New Year's Eve and that sounded like a fantastic idea and I couldn't wait.

**Day 1**. With a good night's sleep, the team left Geelong on the 30th December 1998 and we started our epic adventure north.

It wasn't long before the banter started on the CB radios, which were fitted in all the vehicles, with Bryan kicking it off with details of England's win in the cricket. We were lapping it up with our car to car communication and even started a long rendition of 'Alec Stewarts Barmy Army' and after about five minutes my S-Type came to a halt and I was bundled out of the car which then sped off at a high rate of knots.

I was left in the searing heat for about ten minutes before they returned and when I was allowed back in the car it was under strict instructions that I dropped the barmy army chants. I kid you not, these Ozzies don't like it up em! Unfortunately my euphoria was going to be short lived.

The run up through to Port Augusta wasn't too bad and we achieved 638 miles for our first day.

**Day 2**. Our second day would be far more interesting though as we headed out into the open desert through Coober Pedy and onto Alice Springs where the forecast was for some seriously hot weather. On the run to Coober Pedy it was around 37°C, but on leaving, after a fuel stop, to continue on the Stuart highway north to Alice Springs the temp had already risen to an even hotter 40°C.

Until you do a drive like this you will never know what it's really like to be in the middle of nowhere with just nothing for hundreds and hundreds of miles. All you can see is the flora and fauna of the Northern Territories and its vast sand and rock covered desert. It's a place to take testing very seriously because just one small mistake could put you in a whole world of trouble should you break down or at the worst have a major accident. We did carry medical supplies, plenty of food and water and each car had a CB radio fitted but our only lifeline to the outside world in this region was a satellite phone, because your normal mobiles lost coverage just outside of Adelaide. With Bryans' great planning, good drivers and great cars we would hopefully not require the satellite services!

So far the cars were running well with no issues at all to report home and over the next stretch of the desert we would be pushing them hard and driving up to 150mph. Going ballistic up the highway at these speeds takes some concentration and you have to keep your eyes open looking for kangaroos and other wildlife that might be running onto the road. I had expected to see lots of kangaroos but what I didn't expect to see were loads of feral camels roaming around the place. When I first noticed them, I was straight on the CB to Bryan screaming that I must be hallucinating as I can see a line of camels in the heat haze. He was laughing and also telling me to slow down a bit until we had passed them and sure enough there they were as large as life just plodding alongside the highway. I hadn't realised that Australia had camels, but wise old Bryan tells me that there are about a million of them in this area and that they originally came from British India with the early explorers who then just let them go after their trips. The population of camels was at such high levels that the government were planning to do a cull in the next few years.

We continued at high speed and were a little more wary of these large animals. With my cruise control set the S-Type just thundered on through the desert, and as the roads were just so straight I had time to focus on all the gauges, and it's the first time in my life I could actually watch the fuel gauge move. I noted this with Bryan who was then checking distances to the next petrol station and came back on the CB radio to tell me that we had to slow down a bit or we would be running out of fuel.

For the last thousand miles we had hardly seen any traffic apart from the odd road train but as we were closing in on Uluru I had to blink and rub my eyes a bit. At our planned fuel stop parked up by the roadhouse was a London Red Bus complete with all its original markings. I couldn't believe it and I also couldn't miss the photographic opportunity and we swung the S-Type up alongside. It turned out that it was a tour bus taking people to see Ayres Rock who were equally as interested in seeing the S-Type.

Bryan thought it was hilarious and we all climbed aboard to have a very British moment!

**Day 3**. With full tanks again we continued our final leg to Alice Springs. We had completed 763 miles for the day and a total of 1401 miles for the trip already.

With all that we had done since Boxing Day it was not surprising that we had all forgotten that it was in fact New Year's Eve as we rolled into the city limits of Alice Springs. You could see the town in the distance but the first thing I noticed on the drive in were burnt out cars scattered around the place and I questioned Bryan as to whether or not we were in the right place and he confirmed that it's just one of those things.

We arrived at the hotel and parked up the very hot Jags which I'm sure were looking forward to a rest.

The hotel was an Oasis in the desert, and we were all on a mission to party for the night once Mark and Russ had arrived. Unfortunately, they were driving the backup Ford Utility vehicle which was not as fast as the Jags so they just maintained their own pace and caught up with us when they could. It meant that at all the fuel and photo stops the team in the S-Types would get maybe twenty minutes, but the boys only ever got about five minutes. It didn't matter because they were having a blast. Occasionally we allowed them a head start in the morning just so we could catch them up and keep them closer at the next stop.

When they finally arrived we had the most amazing team meal with our aboriginal hosts, and it was the most perfect setting for New Year's Eve. We had an open fire, loads of wine and great company and both Bryan and I were in our element recalling our testing tales from around the globe which the locals and the team also enjoyed.

Whilst we were discussing the history of the place I asked one of the elders why there were so many cars burnt out and he started to tell me about how the government gave the aborigines both housing and vehicle assistance and they were allowed to have a second hand car whenever they needed one. When the car couldn't be used any longer they burnt it. If it was a good car to them, they rolled it over and burnt it allowing its soul to go to heaven. Any car found burnt-out on its wheels had its soul sent to hell. Now I'm not sure if this fella was just telling me a yarn or not but it was a great story and I liked it.

As we sat there in the heat of the evening, embers from the fire twinkling in the star filled sky, we sipped more red wine and enjoyed each other's company until we all passed out well before midnight. I guess the rigours of the hectic long days at the wheel had caught up with us and it was Bryan that woke us all up at about 1:30am and suggested we should retire to bed. He was definitely a gentleman because if this had happened on some of my previous trips you would have been left out for the night that's for sure.

Reluctantly we all got up the next morning knowing that our plan was to back track to Ayres Rock but at least the big cats were already fuelled up and ready to go so we headed out from the hotel compound to what I can honestly say looked like the scene from a war movie. There were bodies scattered all over the place. They were in the side of the road, on the verge, on lawns and benches. My god, what sort of party did they have last night? Bryan is quick to tell me that the aborigines have a genetic intolerance to alcohol and unfortunately even after just a few beers they tend to just pass out and sleep wherever they like. I look over to the dry riverbed and see about fifteen to twenty of them sleeping by the bank. It a good job that they live out here because if you did that in that in the UK it would get very messy and very wet!

So, our New Year's Day would be a day trip to Ayres Rock. A 654-mile round trip for a quick photo and have the bragging rights of 'Been There, Seen It and Got the T Shirt'.

As usual the traffic situation was chaos, I think I saw about two cars in 300 miles, but I guess it was New Year's Day after all. As we arrived there were actually a few tourists dotted around and they were quite keen for some S-Type pics as we parked up which is always nice to see.

It was still very hot and set against a gorgeous blue sky the rock looked amazing. I would have loved a closer look but with little time we did our brief photo op and turned around in the visitor's car park and sprinted back to Alice Springs. When you put it into perspective that's like driving from London to Scotland to take a scenic picture of Hadrian's Wall, have a quick cup of tea and then drive all the way back in the same day except that we managed the distance about four hours quicker as we averaged 79mph for the whole day to reach a total of 2055 miles and only 7945 to go!

Having had a very late New Year's Eve we decided on an early to bed, early to rise strategy for our longest leg of the trip.

**Day 4**. It was going to be a long one but at least the fifth match of the test series was starting in Sydney and an England win would tie the series. Bryan and I started the day with some early morning banter with our Ozzie contingent and we even gave them a fantastic rendition of the National Anthem for good measure.

Our flight path today, traffic pending, would take us through some unusual places with names like Ti Tree, Wauchope, Tennant Creek, Daly Waters, Katherine and Noonama and we would see a massive change in the scenery with lots of green trees and a more tropical look with high humidity that would give the aircon systems a good workout keeping us nice and cool.

It was a very good day for durability mileage with the trip counter showing me 944 miles for the day and the total was now at 2999miles or basically Lands' End to John O'Groats, back to Lands' End and then back to John O'Groats again. This place is massive by the way.

Our stay in Darwin did give me a chance to hook up the computer and transfer all the data back to the development teams and let them know we were all still alive and so were the cars. Nothing too serious to report back other than specific heat related concerns that didn't bother the team too much.

The long day had again taken its toll and the whole team were straight into bed on the agreement that tomorrow night would be a no excuses team meal and beers at a local restaurant.

**Day 5**. As drives go out here it was just your typical average Sunday drive out to Timber Creek for some tucker in an 815 mile round trip that would take us to a total of 3814miles. The only drama of the day was that England were not doing so well in Sydney and Bryan and I decided to keep a low profile and pretend that we were not that interested. We focused on the driving but fatigue was starting to get to us just like our cricketers so we imposed a max stint at the wheel of 1 hour and with three of you in the car you then got an hour as co-pilot and an hour in the back for nap attacks and recharge your batteries.

**Day 6**. Another little day trip out back tracking to a small town called Daly Waters or as Bryan put it 'A watering hole just a little bit north of the septic stump'. It was to be fair a decent and very welcoming watering hole in the outback and the drinks were refreshingly cold.

The so called town had a population of about 10 and was given its name by the explorer John McDouall Stuart, whose surname was given to the highway we were driving on after he made 3 attempts to travel from the south to the north in the 1860's. He came upon a series of natural springs and named them after the Governor Sir Dominick Daly.

When I checked into the history of Mr Stuart I found it quite amazing that his third attempt from Adelaide to Daly Waters in October 1861 took nearly seven months and he continued to

Darwin in another two months and there we were doing the same route a hundred and thirty eight years later in about four days and that's what you call progress.

Anyway, the Nano town did have a very famous place called the Daly Waters Pub and it was only right that we ventured in to take a look. We parked up the cars in a prime location and wandered into the most amazing room. On every inch of ceiling and wall there we bra's and pants stuck to it from all over the world. From what we were told this all started as a drinking bet in the 80's between a coach driver and his female passengers.

We had a very rapid lunch and said farewell to the owner but on leaving the bar area we went outside to see that our two Jags had attracted some attention but one fella stood out with your typical Ozzie hat with corks wearing a vest and holding a can of beer. The cars were parked next to a massive promotional board for Castlemaine XXXX and with this fella stood there looking puzzled by the new steeds it made a brilliant photo.
Bryan had his camera out and ready and managed to catch the shot perfectly. After we had finished the test trip Autocar used the pic in the magazine article with credit to Bryan and I know it was a very proud moment for him.
After the pictures were finished, I had a quick word with the star of the moment who turned out to be a local farmer who was out ranching his cattle but there were no horses for this chap. His land was about the size of Warwickshire and the only way to herd cattle was by helicopter. With this information I looked from him to his beer and back again then he grinned and said, "No worries mate I'm not flying today."              '

We wrapped up our stop and went straight back to Darwin with a short day and only 748 miles covered. When I mentioned it to the team you could see the disappointment in their faces as we were all pushing to get the cars to the required mileage for our return to Geelong but I could also see that they needed a rest and a couple of early nights so I agreed with Bryan that we may stick to the main route and cut back on the additional mileage loops and just see where we get to on our return. The total for the trip was now at 4562 miles and only 5438 miles to go.

**Day 7**. This was to be our final day of driving out from Darwin with a nice fast loop to Top Springs through some very nice sounding places like Adelaide River, Pine Creek, Katherine, Willeroo & Noonama. Our progress though was slowed down a bit with some tropical interludes coming out of Darwin but as soon as we reached the desert we were back on track and making some great headway, which is more than could be said for the England cricket team. Having left themselves a chance to win the match they were making a right pigs ear of it and Bryan and I decided that we should both be in the same car to listen in to a very patchy radio service.

One of the day's highlights that made the team laugh was when I went past a ring-necked lizard at high speed. I saw this fella stand up on his back legs and make his fan stick out around his head. I assume he was making a statement to the oncoming Jaguar to say back off, but we didn't and as we flew past the poor thing spun around like a spinning top a few times and landed on his back. As I go onto the CB screaming to the others "Did you see that big lizard" they were already laughing when they replied.

We returned to Darwin one more time having completed 907 miles to a total of 5469 miles. An amazing amount of mileage covered and the team were very happy although not the same could be said for the two Brits. England had managed to throw their wickets away within four days and Australia won the Test Series. With our glum faces we decided that the show must go on and what better than a good team meal and a few beers at a restaurant that Bryan liked the look of. I asked him what he suggested and all I got back was "Mr Moore, please put your digestive organs in my hands" I knew I was in for trouble and I wasn't wrong. My tasting evening had me eating Alligator Bites, Camel, Kangaroo Tail and some fish called a Barramundi.

Watching the team bond and enjoy themselves was great and Bryan was taking audience at the head of the table just like a ship's captain. It sat well with him because he did look like a seafaring type of chap in the style of Captain Birdseye and I just had to tell him that. He burst out laughing and said that he was not surprised at my comparison because he was actually related to Edward Smith who captained the Titanic. It was a wow factor moment that I never checked up but who cares. It was a brilliant moment and the team had a good laugh on the back of it.

Its nights like this that make working at Jaguar so special. A group of engineers brought together to carry out some specific testing in what is considered to be one of the harshest environments on the planet, enjoying some good wine, great food and brilliant company. Happy days indeed.

**Day 8**. The day should have seen us heading east to do the return leg down the gold coast but due to some really bad weather and serious flooding we decided that the best course of action was to return to Geelong down the same road we came in on. We did have some more amazing place names to fly through such as Jabiru, Mary River Roadhouse, Mataranka and ending up at Tennant Creek.

On leaving the tropical weather of Darwin it wasn't long before the familiar dry air was back, and the temperatures rose back to 37°C. Everything was going well until we lost the clutch pedal on the V6 S-Type. Luckily, we were quite close to a roadhouse and we all pulled in to take a look. No sooner had we pulled up and Mark shouted fire and as quick as a flash the team had a fire extinguisher out and the thermal event was no longer.

It turned out that we had taken some damage to the clutch pipe which then dumped a load of fluid on an extremely hot exhaust. The boys decided that we couldn't do anything about it and we really needed a workshop to fix it but the nearest place to do this was about 300 miles away. Fortunately, the Stuart Highway is straight with little traffic so it made driving the car quite easy and I ended up driving the full distance to Tennants Creek without once having to stop. Let me just say that again…. nearly 300 miles without having to stop. Try that on the M1 if you dare (no seriously don't you will get arrested).

Whilst we were doing this, we also managed to hook up the satellite phone to get assistance. Luckily, we managed to get hold of George Hall who was working in the USA for the S-Type team alongside the Lincoln team and he then managed to hook up with a local Ford garage near the Creek. Our ETA into the Creek was going to be about 10pm and the owner left the keys hidden for us with the instruction to use his facilities and do what we needed to do. When we were finished, we could put the keys back and get to the hotel for some rest. Being part of Ford certainly helped us that day!!

We rolled into Tennants Creek and Bryan and I decided to stay and help Mark and Russ with the V6 allowing the others to find the hotel and check in.

There was no spare ramp but there was an inspection pit, so we rolled the car over and allowed it to cool for a bit while Mark went out to find some cold beers. Our luck was in as the fuel stop had beer and a box of ice to put them in, so all was good in the world and we headed into the depths of the pit.

The lighting was poor, but Bryan managed to locate a large inspection lamp and we plugged it in. It lit up the car and the surrounding area like the lights at Wembley and before we could start our investigations we were joined by every bug and insect in Australia. It was carnage as we swatted and stomped on anything that looked like it was going to bite us. The Ozzies took it in their stride but Bryan and I were struggling to help so we spent most of our time on bug watch. It took the guys about two hours to fix it and by now our insect audience was huge.

We locked up and made our way to the hotel for another cold beer and into bed at 1am having completed 748 miles, put out a fire, fixed a clutch and made friends with a million or so bugs.

**Day 9**. Well we started in the vehicle pit then moved to my sleeping pit where I was rudely interrupted at 6am, having had five hours sleep, by a massive lizard on my headboard. I leapt out of my bed to fetch assistance and the hotel receptionist calmly came into the room and picked it up and gave me the 'you stupid boy' look before taking it back outside.

I swear it was the biggest lizard you have ever seen so I'm not sure why she tutted and gave me the rolling eyeball look.

I was now in the wide-awake club so it was only fair that I wake everybody else up for an early breakfast and early start for our next leg to Coober Pedy which would return us 758 miles to a total grand total of 6975miles.

It's a small town with less than 2000 people who live there and it's a strange old place that owes its existence to the opal trade and lays claim to being the opal capital of the world. The bit that makes it strange is that most of its inhabitants live underground in places called 'dugouts' which are cooler to live in as the daytime temperatures are scorching hot.

As we were intending on staying there for the night Bryan had booked us into an underground hotel that started its life as an opal mine. The original boring tool was used to go back down the main shaft and turned through 90 degrees to create individual rooms and I have to say they were both cool and cool if you know what I mean. I managed to have a great night's sleep in there but when you wake up there is a slight amount of reddish dust on everything from the bare rock walls and ceiling.

Whilst the town is a bit strange it does have the most interesting golf course in the world. Due to the daytime heat, which is normally in excess of 100°F,  games are played at night with glow in the dark golf balls and as there is not a blade of grass to be seen in the desert landscape each player carries around a small piece of turf to use at each tee. Completely bonkers on all levels. They told me that Billy Connelly played there once during filming for one of his TV series.

Even though our stay is brief we manage to find out that the film Mad Max 3 was filmed there and some of the props and cars were left for tourists to see. The town was also featured as one of the stop offs for another film The Adventures of Priscilla Queen of the Desert.

**Day 10**. A slow day that saw us complete only 655 miles to take the total to 7630 miles as a result of the imposed speed limits after you leave the Territories and we also lost some time for a great photo stop by a salt lake that nearly lost us two expensive prototype cars while doing it.

Bryan had been telling me all the way down about this great place where the Darwin to Adelaide train line passes the salt lake and that it would make a brilliant photo opportunity, so who was I to argue.

Turning off the Stuart Highway we went east towards Lake Eyre and located the spot that Bryn had told us about. It did look good and Bryan wanted both S-Types either side of the track. As we were setting it up I mentioned to Bryan that the cars were a bit close to the track and looked a bit dangerous should a train come along. He started laughing and told me that the main line is for The Ghan train that passes just once a week and the chances of it passing today were zero.

The sound of the horn from the train some fifteen minutes later stopped us dead in our tracks for a few seconds. Franticly the team rushed around to move camera equipment and tools out of the way, but it was too late to move the cars.

We all closed our eyes as it thundered on through giving another blast of the horn. We stood there frozen as this enormous train, with an endless line or carriages pounded the rails and eventually faded away on its way to Adelaide.

There was a nervous silence for a few seconds quickly followed by nervous giggles that developed into full blown belly laughter. With inches to spare both cars were unharmed, and you couldn't help but see the funny side and I can't even think about the conversation back to the UK if they had have been hit. I spoke to Bryan many times after this and it always made us laugh. I also contacted Mark Rowlands to find out what was the best bit of the trip and his response was also 'The Train Mate.'

**Day 11**. The last leg of our epic journey was a run back into Geelong covering another 680 miles to our total of 8310 miles and just 1690 miles short of our target.

After the events of the previous ten days the last stint was quiet, and I wasn't surprised that tiredness was setting in. During one of my passenger stints I had time to reflect on what we had just achieved. We had completed an epic journey through the central part of Australia from the South to the North, accumulated three days of high-speed mileage out of Darwin before returning down the same route. The vast majority of the trip was made possible by the explorer John McDouall Stuart who traversed the outback in 1861 & 1862. It took him three attempts to complete the route taking nine months in total to complete just one way and here we were having done the same journey and then returning in just eleven days. With no speed limits outside of the towns we had covered hundreds of miles with the speedo sitting well above a hundred miles per hour. In the V8 I completed a whole stint at the wheel with the cruise control set at Vmax doing 149mph until my fuel ran out. On a nine hundred- and seven-mile leg we achieved an average speed of eighty six miles per hour which is bonkers. In doing this we had used just under four thousand litres of fuel and apart from the clutch incident both cars behaved brilliantly, and the team were very impressed making my final report back to the UK an easy one to write. There were a few hot climate niggles that were reported back and the team in the UK were quick to fix these before the cars went into final production.

None of this would have been possible though without the help and dedication of all the team who gave up their Christmas holidays to be part of something special that went into the history books at both Jaguar and Ford.

To honour the occasion Mark decided that we should all go round to his place for an Australian style BBQ with some great food and lots of Victoria Bitter. We sat all night long talking about cars and the epic trip we had just been through. The boys were telling me that no other test trip had been quite like this one and certainly nowhere near as fast. They were thrilled with the S-Type at just how robust it was in the conditions and constant high speeds and were even more thrilled with the overriding sense of achievement although it wasn't all over just yet!!!

Nursing a thick head from the Victoria Bitter I sat in my room contemplating my return to the UK when I took a call from the head of Public Relations Martin Broomer. He had been following our road trip with interest and had been discussing the possibility of doing an article with Autocar magazine as part of the final validation of the car.

He then dropped the bombshell, "do you fancy staying out there to do a drive with a journalist?"

It was an easy decision to make really as I wanted to stay and get the car to 10,000 miles and complete the target given to us. Martin advised me that Autocar were sending a bloke called Hugo Andreae who would be with us in a couple of days, and if possible, could I work out a good drive route and also find some places of interest for some great pictures.

The cars were taken down to the proving ground for a couple of days' worth of checks and I went back to Geelong to sit and work out a drive route with Bryan. We decided the best bet was to do a Melbourne to Sydney route via Corryong, Thredbo Village, Cooma, Narooma & Kangaroo Valley and then ending at the iconic landmarks of the Opera House and Sydney Bridge.

As we sat looking at the route we realised that the mileage would get us close to the 10,000 miles target that I wanted to achieve when we first set out.

With the route in the bag and some places of interest sorted Bryan advised me to take a look at the Great Ocean Road as a start point so I set off for a bit of discovery.

The Great Ocean Road was built by ex-servicemen returning from WW1 and was built in honour of those who lost their lives. It stretches for around 150 miles on the south east coast between Torquay and Allansford and is the world's largest war memorial. It really is the most gorgeous road and it's a shame that the speed limits prevent you from enjoying it even more. The police in Australia view speeding as the most heinous crime that you can commit and I'm sure that 5mph over the limit would get you a public flogging of the highest order so with that in mind I drift along the road just enjoying the view.

My first stop was at a golf course that was covered with kangaroos and as I had not been up close to one yet I ventured out expecting to find Skippy and get myself some great pics. I ended up finding out that they are not quite as friendly as Skippy and my approach for a bit of fuss was met with a dog like growl. I left them alone for today and continued north until I found the most amazing place called the Twelve Apostles which are the most magnificent rock stacks that you could ever see. They rise from the sea and stand about 120ft tall and were created by the constant erosion of the limestone cliffs millions of years ago. They then formed caves and they eventually turned into arches which collapsed to leave the stacks. One of its most famous arches was called London Bridge which collapsed in 1990 leaving two people stranded on the stack. They were there for three hours before they were rescued by helicopter.

Anyway, they looked stunning and I continued up the road a bit for a better view and came across an airfield with helicopter rides. I went in to take a look and mentioned that I worked for Jaguar Cars and we might be looking for some aerial photography of our S-Type driving up the coast road and they were more than keen to help. To my surprise the pilot suggested that as they were not busy why didn't I have a quick flight to take a look.
So, in the interests of preparation for PR filming and for the love of the company I accepted the offer and strapped myself in. It was an old-fashioned helicopter with the full glass bubble at the front which allows you to get a far better view.

We took off and followed the road back to where I had come from and it looked amazing. It was only then that you appreciated just how much hard work went into making it. The long lengths of tarmac wrapped itself around the coast line like an enormous snake. We turned back and made our way to the Apostles and the pilot suggested that the best way to see them was from the sea looking back towards the coast, so we flew out a bit and swung back in. We were skirting the ocean and he asked me what I thought. I mentioned that they really are magnificent, and he said "you aint see nothing yet."
As we approached the stacks at some speed, he pulled up the nose and we climbed out above the tallest stack and pivoting around the main rotor he pointed the helicopter back down towards the sea. Coming back down the stack he pulled up and

147

we banked back over the top towards the airfield. It was a good job I had nothing to eat because if I had I'm sure it would have been ejected!! It was very impressive though and one of the highlights of the trip.

For the next part of the adventure we collected Hugo and his photographer, Warwick Kent, from Melbourne airport and headed off for a round trip up the Great Ocean road to get all the glamour shots out of the way first and then headed back for a good night's sleep.

The next day the three of us along with Mark Rowlands had our sights set on the coastal resort of Merimbula via some inland excursions to more photographic opportunities and a drive over the Snowy Mountains.
We travel for hours with hardly any input into the steering wheel before finally arriving at the Hume Weir Reservoir and it was well worth a visit. The intense heat of the summer had shrunk the size of the lake and exposed the bleached trunks of old dead trees. It was too good to miss for Warwick and he got me to put the V8 as close to the water's edge as possible for another brilliant photo.

We continue to the Snowy Mountains and the temperature drops from high 90's to low 60's as we climb up the road.
We pass through the skiing village of Thredbo which was the scene of a terrible landslide in 1997 killing 18 people. Until I read this info on a local board I didn't even realise that Australia had ski resorts as I only ever saw or learnt about the hot climate side of the country.
As we leave the village the east coast weather changes quickly and we are suddenly in thick fog which changes to light rain quickly followed by some of the heaviest rain I have ever been in. The deserts of the Northern Territories seem like a long time ago now as we arrive at our slightly cooler overnight stay.

Our final day to Sydney starts early as we would like to get into town and see some of the highlights before it gets too dark. Hugo is enjoying the drive and seems very keen on the S-Type and Warwick is getting hundreds of pics for the magazine so all in all things are going very well. We push through Kangaroo Valley and some three hours later we are in Sydney rolling up to our final destination at the Ritz Carlton hotel. As we have a journalist with us it's nothing but the best thanks to our brilliant PR team. I tell Mark that if it was just you and me, we would be in a motel just outside of town mate so make sure you enjoy it!

As we go to park the vehicles, we note that the odo has just tripped over 10,000 miles and I'm delighted to have reached our target mileage. With beaming smiles, we walk to the reception and I'm greeted by a very well dressed chap who tells me that I have just ruined his day. It turned out that he had ordered a new BMW and he now wished he had waited. You couldn't have planned that moment and Hugo was quick to note it for the magazine.

We headed up to our rooms and the porter dropped my bags off in the biggest and most unbelievable room I had ever stayed in. I could not believe this was for me and then I thought that maybe our guest journalist had got my room by mistake and I was in his. I just had to ring him and ask if I could pop down for a moment.

I walked into his equally if not slightly better room and was very excited that no mistake had been made. We had a laugh when I told him why I was checking out his room and I went back with the biggest smile on my face.

The smile only got bigger when the porter told me that Madonna had once stayed in my room.

It was a brilliant end to the most amazing drive you could ever do, and it will never be beaten.

With the change in speed limits imposed in 2006 it would be highly illegal to do it that fast ever again, but it was more than that. We had started and completed a validation test of 10,000 miles which was just the beginning of a 100,000-mile test that the cars went on to achieve at the proving grounds in Geelong and enabled the S-Type to be launched with confidence. It had also brought a team of engineers together to form great friendships and create great memories.

Unfortunately it is with great sadness that I finish this chapter having just learnt that Bryan Knowles, who was the Vehicle Engineering legend from Geelong, passed away in June 2018. I emailed him to let him know that I was finishing my book and I got a reply from his wife Jo who said that my email arrived on what would have been his 72nd birthday.

Without Bryans support, knowledge and dry humour this epic trip would not have been as good as it was. Nothing was too much trouble and his energy and enthusiasm were intoxicating.

I will always remember the look in his eyes when the Ghan train nearly hit our cars followed by the uncontrolled laughter afterwards.

I dedicate this chapter to you my friend, because I could not have done it without you. We had a blast mate!

# 13. The Need for Speed in Bonneville

For as long as I can remember I have always had a fascination with speed and the desire to be fast.

Even when I had a wooden go-kart I was always trying to make it faster by pushing harder or making it lighter or more aerodynamic. I had my first small motorbike when I was quite small but no sooner had I started to ride it and I wanted something bigger and faster. I'm not sure where I get this desire from because my dad was a polar opposite. Our drives to Great Yarmouth every year were taken at precisely the speed limit and it took us hours. As cars went past us doing just ten or fifteen mph quicker he would lambast them for being hooligans and I just sat in the back wishing I was in the other cars. My Uncle Ray on the other hand was a speed freak and I think that must have rubbed off on me a bit because he actually built me my go-kart, built my first motorbike, took me bike racing and then introduced me to flying so I have to dedicate this chapter to you!

Throughout the XF project we had built up a great relationship with the PR team and they really liked the way the Vehicle Office did business. We worked very hard, but we had so much fun doing it. It was during one of our PR discussions that they asked us for ideas on what we could do with the new car to create some media attention that was a little bit different. There were quite a few ideas that were worth a shout but I had just one thing on my mind, and that was to set a top speed at Bonneville. During testing I had a few runs with the unlocked calibrations and the car was extremely fast and my engine build team lead by Dave Warner and Simon Stacey were convinced it could do a lot more. With the data in my back pocket I suggested the idea to the team and they loved it and off the back of this we could launch the car at the LA Motor show.

With the thumbs up from my Engineering Senior Manager Andy Whyman I did what I have always done and threw myself into the project like a man possessed. The first thing on the agenda was to secure a suitable vehicle from the engineering test fleet. Not too difficult a task seeing as we owned the fleet. With the donor selected the next job was to convince the prototype operations team that they needed to build it. I went over to see my old mate Adi Smith and as soon as I mentioned Bonneville he was in, hook line and sinker. Within a few hours we had a build plan and a designated area in the workshop to build it. That's the great thing about the guys at Jaguar, they were always passionate about the company and when you were doing something to promote it they were so supportive. Unfortunately this passion has dwindled over the years and I think it's because people are just more focused on their own ambitions, goals and objectives. The management when I was doing all this crazy stuff were born and bred Jaguar who were quite often second or third generation employees that quite literally lived for the brand.

Luckily the powertrain team still had that passion and were fully on board to build us the best engine possible using the AJ133 V8 as the base engine. Dave and Simon had a few ideas with the supercharger and were confident of building us a great engine for the task at hand.

With the wheels in motion I then set up some discussions with our PR team to work out how we were going to do the actual speed runs. I was hoping to get the nod from Ken McConomy to fly out and do the runs myself, but Jaguar had connections with a team already out in the USA racing the XK8 called Rocketsports Racing who were more than keen to help out. As the project leader I was given the task to contact the team and come up with a joint plan of attack that would see us build the car, test it and then fly it out to the USA for Rocketsports to finish off and prep it for the salt flats.

The head of RSR was a chap called Paul Gentilozzi who made his name in Trans Am racing and had a lot of success with Jaguar in the past. His current team were running the normally aspirated XK8's over in the USA against Corvette and Porsche. I set up a number of dial in meetings with Paul and his team to discuss the build of the car and making sure the basics would conform to the race specs required for Bonneville. We got as much weight out of the car first getting rid of anything that we didn't really need so just about all of the trim was ripped out and recycled. We replaced the standard seats with bucket race seats and then set about designing a roll cage for the car.

Adi was loving the build and although we were trying to keep it secret word got out at Whitey and we had a few keen visitors taking a peep at what we were doing. The interest is great though and anybody who took the time to come down was always given time to take a look at the build.

Progress was going well but the main part of the car was just about ready. Dave and Simon had taken a stock engine and transformed it with a lot of experience, shed loads of black magic and just a hint of pixie dust into a 600hp monster. It was dropped into the car and connected with a lot of excitement from all of the team and it wasn't long before we could crank it up and make sure it all worked. With a straight through exhaust system it was guaranteed to make a noise, but we didn't expect the thunder it delivered.

We managed to get the whole team around the car for its first start and she didn't let us down. On pressing the start button to get power all the gauges did their start up sweep that always looked cool and we were ready for ignition.
On the second press her heart burst into life and the super unleaded fuel flowed through her veins while the supercharger scavenged for air like a newborn child's first breath. The roar from the exhaust made the hair on your neck stand up and the crowded workshop loved it as we revved it up some more.

Our dial in with Paul and RSR that evening was a little more exciting now that our baby was alive and we even played a sound clip over the phone to prove the point but our excitement was short lived as Paul advised us that due to it being the end of season at Bonneville and with the weather changing our window of opportunity was getting smaller and we really needed to get the car out asap for it to be prepared in Michigan and then trucked down to the salt flats.

My team were fully on board and worked all the hours we could with some sleepless nights thrown in for good measure, but the clock really was ticking.

With her flight booked there was no turning back and we just had to deliver. I got a call from Paul to say that the one thing that concerned him was that the new and unlocked engine calibrations really worked because once over in the USA all they could do was add the aero package and some more race magic before the car had to be shipped.

What he confirmed was that under no circumstance could the car go on the plane without confirmation that it would go over 185mph which put the pressure on us a bit with less than 24hours to go.


The next morning, I got together with the build team to confirm that she was checked and ready to go but the weather was not looking good. I called Millbrook which was my first choice and they sadly informed me that it was raining and I got the same response from Bruntingthorpe so it wasn't looking great, but all was not lost.

It's at times like this that you have to go the extra mile and I was presented with an opportunity that allowed me to drive the XFR in excess of 185mph confirming the calibration changes were working. To all those involved that made it work I thank you for that ultra-top secret test session. As I hit nearly 190mph I had the biggest grin ever!


Our baby made the flight and arrived in Michigan on time and the RSR boys set about making some significant updates to improve the aerodynamics and get us a better top speed. All the gaps in the bodywork were sealed as was the main radiator grille, rails were added to the roof to make sure all the air went back towards its new and very large rear wing. An all-aluminium underfloor was added to make the air flow really smooth and a set of smaller cross section tyres were fitted to also improve its resistance to air flow.

Some extra magic was added to the front grille with some air intake pipes creating a ram air effect and then the piece de la

154

resistance was some nitrous oxide to feed into the intercoolers to ensure they worked at their max by keeping them very cold. In the event that we reached a significant top speed and required some assistance with stopping her in a short distance a parachute system was fitted at the back just for good measure.

So, on November the 7th 2008 our pride and joy took to the salt flats with Paul Gentilozzi at the wheel.
The first runs began and the car suffered trying to keep traction at high speeds as we were using Michelin Pilot Sport tyres inflated to 50psi to improve rigidity and reduce drag. The problem was resolved when Paul suggested adding ballast to the boot with an extra 250lbs to make it sit down

a bit. Finally the team set off to see what she could do and with the rev counter showing 6250rpm our XFR had peaked at 225.675mph even though we were still suffering from wheel slip on the salt.

As the times and speed came through we were both ecstatic and emotional. It was a hell of an achievement by a small and dedicated team of Jaguar petrol heads going that extra mile. What we had just done was break Jaguar's long-standing speed record held by Martin Brundle in the XJ220 at Nardo in 1992 when he reached a timed lap at 217.1mph. It really was epic and after the event we realised it was even more impressive because the salt flats are at an altitude of 4200ft and the salt surface generated a higher drag factor than what you would encounter on tarmac, so in reality our 225.675 was probably nearer to 250mph if we had been on tarmac at sea level.

What an achievement! We had produced the fastest ever Jaguar, and I need to point out at this stage that it was the fastest ever Jaguar on four wheels as my friends in the RAF would soon be chirping up a bit if I didn't.

The PR team were thrilled, and the Media wheels were rolling. The car made its way to LA for the show and the rest as they say is history.

Our baby returned to the UK and I drove her at the Reims racetrack before she was retired to the Jaguar Daimler Heritage Centre at Gaydon where you can still see her today.

Well done to all the team for a great achievement.
Adi and Mick from Proto Build, Dave & Simon for the engine, Ian Parsons and Gary Embleton for the calibration support, Roger Wardle for breaking all the rules with the ZF transmission, Kev Gaskins for a funky cooling system, Dan Banks for being an Electrical Wizard (they would have burnt you at the stake in the old days), Andy Whyman and the XF 'A' Team for being Vehicle Engineering Legends and finally to Paul and the RSR team for putting the cherry on top of a very fast cake.

All of the above have become my best mates and we all still talk about this with so much pride and passion and I'm sure Sir William Lyons would have approved but the special thanks go to my Uncle Ray who inspired me to go fast or go home.

# 14. Chilling with Jay Leno

So, there I am in Phoenix doing the world press launch of the all new XF and I'm helping set up about forty XF's for the journalists to assess over the next couple of weeks. We have established a great base camp in the car park to some swanky hotel for the rich and famous, the cars are looking amazing and the plan is coming together for an epic road drive that hopefully will generate loads of column inches in all the motor magazines.

We are a couple of days into the prep when the PR Team ask me if I could send one of the XFR's to LA for a photoshoot and also be available to fly out and provide technical support for the day. It's a daft question really, I mean, who wouldn't want to fly out to LA and do a photoshoot? It's then confirmed that I'm spending the day with non-other than Jay Leno at his famous garage. Well, I try to look cool on the outside, but on the inside I'm freaking out like a teenager with tickets to see Britney Spears.

One of the great things about Jaguar is they always have a great PR Team and they really look after you. At the end of my shift I go back to the hotel and there is a letter in my room detailing the trip to LA the next day with flight details and a brief on what to say or not say if on camera. They have also arranged for a chauffeur to collect myself and the PR manager Jonathon Griffiths the next morning. It's just a small detail that make your day. Simple but so effective in making you feel special.

When flying back from Mexico once with a journalist I asked him how well Jaguar performed at drive events and he said that any event with Jaguar was very special. The attention to detail was always first class and the team always made you feel like part of the family before you even got near the car.

The night before our trip to LA I sit up checking out all I can about Jay Leno…sure I know he has a TV show and he has a garage with quite a few cars in it but I want to find out more so that I'm well prepared for the day.

As I read page after page of details, I find out that he fell in love with a Jaguar XK120 when he was a small boy. He wrote that in 1959 he walked up to the top of the hill by his home where he had a moment….an Automotive Epiphany.

There outside a barn being polished was a blue Jaguar XK120. He was beckoned over by the owner who allowed him to sit in the car and it immediately ignited his passion for the brand. The car had quite literally brought him to life.

This was great information that I hoped I could ask him about the next day. I could also easily relate to this because as a young boy I used to spend Sunday mornings watching and helping my neighbour's son Andrew clean his new Ford Escort RS2000 MK1. I loved that car…. Sebring Red, alloy wheels and an engine you could eat your dinner off. I used to drool at the rally versions of the MK1 Escort and dream about having one. It was some years later as a nineteen year old that my dad said that he had been speaking to Andrew in town and now that he had a family he was thinking about selling his RS2000……the original car I had drooled over every Sunday. Well my dad very kindly helped me out with a loan and there she is on our drive….the Automotive Love of My Life!

The chauffeur arrives and Jono and I set off for Phoenix Sky Harbour Airport and the short hop to LA.

We touch down at Bob Hope Airport which is only a stone's throw away from Jay's Big Dog garage. As we pull up to the gates there is a big truck waiting for us with our new XFR on board. The truck had left the day before from our base in Phoenix and driven through the night to make sure it was on time. Leaving the guys to unload we wander off to meet the legend.

I was expecting an entourage of TV people, film crews and alike but there was nothing. I asked a chap if he knew where Jay was and he pointed over to a workshop. We strolled in and there by a lathe in a denim shirt, jeans and work boots was Jay Leno who was making some parts for a classic Ferrari he was rebuilding.

Hey guys…. welcome to my shop….lets go over and make some space for this new Jagwaaaar of yours.

We went over to his main garage and as we entered the hanger you just had to stop and take a breath. It was Automotive Porn of the highest order….gleaming cars all over the place to rival any motor museum on the planet.

Jay quickly led us to the far end of the shop to an area for the photo shoot. He referred to it as his Jaguar shrine and proceeded to tell us the story of falling in love with the XK120. Having read it some eight hours ago I was quite familiar with the details but to just listen to the story and the way he told it was fantastic.

An area was cleared, and I brought in my own pride and joy to join the Jaguar shrine.

In no time at all the film crews arrived and Jay was soon in front of camera talking about the XFR and sharing all the technical details we had discussed only five minutes earlier. It was clear that this man had an absolute passion for cars and in between takes he took some time to discuss the XFR some more. I asked him why he thought the Jaguar was such a gorgeous car and his response is one I still use myself today. He said that when you wash a car and it feels like you are running your hand over the curves of a naked woman whilst bathing then you know it's perfect and he is so right! Don't ever get it wrong though and tell the other half that her perfect curves feel just like an XK or XF…. they just wouldn't understand.

With the film crew and photographer filling their boots I managed to get some time with Jay to discuss our love of cars. He didn't realise I was an engineer responsible for the development of the XFR and he seemed quite keen to do a detailed tour of his hanger for me. We strolled and talked for an hour or so and I mentioned my love of two wheels. Hey you're in for a treat……. he went off to get some keys and opened the doors to another part of the workshop that was full of bikes. It was a kid in a candy shop moment….a 1933 Indian, 1930 Brough Superior, 1955 Vincent Black Shadow and an MV Agusta right through to modern day Yamaha's and Kawasaki's. He let me sit on them all and each one came with a new story.

As we discussed all the technical details he had a little surprise up his sleeve. We wandered over to a very sleek black motorbike and he stood back and motioned for me to step forward and take a look. It was awesome. "I bet you don't know what that is" he smirked. He was right…. I had no idea what it was but I did know it was a small jet turbine. It was in fact a small C18 Rolls Royce Allison engine used in the first production models of the Y2K Jet Bikes and this was chassis no 002.

I was so intrigued that he asked if I would like to start it up to see what it was like and if I liked it. Well what's not to like…it's a motorbike and a jet…. the most uber cool thing you could ever own!!

Letting it warm up was just heaven….the smell of avgas and the jet wash at the rear gave you goose bumps.

Jay then told me about a ride on the main street and a car got too close at the lights…. the jet wash melted the plastic bumper right off his car!

We could have talked cars and bikes for hours, but we needed to press on with filming and I had a need for the rest room. Jay directed me to his own private area and said help yourself. I will see you by the car when you are ready.

I enter the bathroom and make myself comfortable on the throne whilst I gather my thoughts about the day so far and what's left to say and do for the PR team. As my mind wanders, I suddenly realise that there are a number of pictures on the wall of the bathroom and all the people in them are looking at me. On closer inspection they are all pictures of Jay shaking hands with Presidents of the USA. I couldn't help but chuckle at the situational humour…. there I am….a Nuneaton lad on the private throne of one of the most famous men in the USA being watched by 5 presidents of the USA.

Still laughing I wandered over to the shrine where the team had the car ready for the drive. A journalist from 'The Times' newspaper joined us and we head off to the hills.

Jay loved every minute of it and before long we were on the freeway overtaking everything.

I was a little concerned about being pulled by the cops but Jay informed me that this was not an issue as he was an honorary member of the LAPD….just how cool is that.

Having had a spirited drive for an hour or so Jay heads off into an area of LA to buy us lunch at his favourite pizza place. The staff treat us like royalty and 'Pizza for Leno' arrives in the blink of an eye.
As the journalist and Jay discuss an article he is doing for The Times on Sunday I'm busy pinching myself under the table.
This is a special day in so many ways and here I am chilling with Jay Leno eating pizza……BONKERS!!

As we tuck into some excellent pizza there is a constant stream of people wanting pictures, selfies and autographs and I'm just amazed at how popular Jay is. As he sits there posing for every photo, I can't help but think the President of the United States would probably get a quieter moment for lunch. Throughout our lives I'm sure we all dream about celebrity and wonder just how cool it might be to have the glamour, the fame and the fortune but I can tell you this, after an hour it all gets a bit much. Sure, it's unreal for me but at least I get to eat some pizza and have a drink without a whole lot of faff to deal with!

Anyway, we head back out and I suggest I drive but the big fella is having none of it and jumps in behind the wheel. He really is a petrol head!!  We hit the freeway and leap into Leno Hyperspace but hey, I'm relaxed now I know he has friends in high places.
We end up back at his hanger and wrap up a fantastic and surreal day. There was just far too much to take in over one day but we said our goodbyes had the obligatory portrait pic with the great man and set off back to Phoenix and the reality of our day job that 24hrs ago seemed so glamorous but was now overshadowed by our LA soiree.

# 15. Top Gear Stuff and the Sweeney

During the later phase of the XF project we were getting requests from the media to feature our new baby and it wasn't long before the Top Gear request landed on the meeting table and our Vehicle Engineering manager Andy Whyman looked over to me and in classic Whyman style said "Your mission should you wish choose it is a trip to Devon to film with Top Gear." Never one to turn down such an opportunity I leapt at the chance to work with the Top Gear team, in fact I had brushed shoulders with its main host Jeremy Clarkson on a few occasions.

The first was back in the 90's when his new Jaguar Saloon arrived at MIRA and I had to put a 1000miles of run in mileage on it and check that it was perfect. He does like his Jaaaaaags and by coincidence my sister was working for British Airways and looked after Clarkson on a flight and mentioned that I ran his car in for him. He said we must have done a good job as it was a great car!

The next time I brushed shoulders with him was out in the desert at Midland/Odessa Texas for an Air show where he was filming for his series of Mean Machines. I was out with my team testing in Pecos with our first supercharged engines and I decided that we should go to the air show at the weekend.

I just happened to spot him walking through the crowds and I stopped to say hello. He was quite surprised to see some Brits and when I said we all worked for Jaguar he was asking us all sorts of questions about what we were testing and where. He even asked why we were so far away from our Phoenix base and I threw him some crumbs about secret testing in the Texas desert. He so wanted to find out what we were doing but we just kept him on the end of the line. He did have an inkling that we were testing the new engine and I just smiled.

When we found out he was filming for a show he invited us along to watch a piece to camera with the Blue Angels display team and it was hilarious. Now the Blue Angels are a damn fine bunch of pilots, but they do put themselves up for some abuse with all the glamour that surrounds them. The spokesman for the team was being interviewed by Clarkson who asked why they had a 747 jet with them. The over keen spokesman responded in a nano second, "Well Mr Clarkson Sir, we have six aircraft, six pilots and nearly a hundred support crew and gear that needs moving from show to show." The response from Clarkson was so funny "Oh, and I thought it was just to carry all your ego's."

The spokesman didn't do banter and just gazed back at him.

A bit later Clarkson was asked if he had enjoyed the show to which he replied, "Yes but it's just not the Red Arrows is it?" the Blue Angels spokesman nearly burst a blood vessel at that one.

Like I said, the Blue Angels are just awesome but they need to work on taking some banter!

Just before we left I shook his hand and thanked him and said we had to get back for testing tomorrow. He asked if the test cars were far away. I said no, as a matter of fact we have one here today in camouflage so see you later.

We did make a run for it so he couldn't see it though.

The next brief meeting was at a motor show and he remembered our chat in the desert and I confirmed that it was the V8 Supercharged engine on hot and high-speed testing that we had with us that day.

The only other meeting I had with Clarkson was at a charity go-kart race that he was hosting to support the McMillan Cancer Trust. It was during our brief time in Formula 1 and the team had been approached to attend along with all the others and a few rally teams. In order to show some connection between the F1 team and Jaguar Engineering they were looking for somebody to drive in the team and fortunately for me the lovely Carol Mason, who was doing the PR, called me and said "hey, you can drive a go-kart can't you?" and the next thing I knew I was being fitted out for a Jaguar F1 race suit! When it arrived and I had tried it on the nerves appeared. I mean, I have a swanky race suit and I'm in the F1 team keeping up the Jaguar image. The pressure was on and I was now wishing that I had lost all the overseas testing weight that I had put on over the years. I looked in the mirror and tried to re-package myself, but it didn't matter how much I sucked it up that weight from years of overseas testing was going to make a difference to the karts performance. With the extra pressure put on myself I travelled down to the track with Carol and was introduced to my teammates including Johnny Herbert and Thomas Sheckter as well as a couple of pro karters.

Well, I was certainly out of my depth at this one then and looking around the star-studded event I think I was the only person in the room that I didn't recognise. It was certainly a cool place to be as it was the who's who of the motoring world. There was just about every F1 team there, a couple of rally teams and the cast from Top Gear and 5th Gear and numerous celebs.

As Carol was mixing it with the teams, I went off to look at the computer simulators and bumped into Clarkson once again. My opening line had to be " we must stop meeting like this" and I told him about our history and again he remembered the Texas meeting which was cool but the best thing was that he was packing a few more pounds than I was so I might be in with a chance of not being the slowest!

Johnny Herbert started our team brief and was telling us all to enjoy the race and we must do everything we can to win it. I had to pipe up and mention that a couple of years back a medium had predicted that at some point in my career I was going to win a race for Jaguar and they all laughed. I explained how I ended up having a reading and that so far 90% of what she said had come true. Johnny smiled and said he hoped she was right.

We started the race with Johnny Herbert in the lead only to get a pit penalty that pushed us right back and I thought we had blown it but by some miracle we managed to claw our way back into contention. With both Johnny and Thomas doing an excellent job it was time for me to get ready. With some last-minute tips from our F1 legend about holding my ground and don't let anybody past I went out and gave it the beans for my twenty-minute slot. We were holding 4th position and I was so determined to keep it that way but there was a kart right up my tailpipe giving me all sorts of grief. I threw away all thoughts of trying to stay with the 3$^{rd}$ placed kart and concentrated on my own battle.

I managed to keep him at bay for the last ten minutes of my stint and I headed back into the pits for a changeover.

Mr Herbert was impressed and gave me a nice pat on the back but then I heard a voice in the pit shouting "who was the wide load in the Jaguar kart?" it was none other than Damon Hill who then came over and said "Well done chap" and gave me a bottle of beer! I was completely star struck to be honest and accepted the beer with a huge smile and told him I was just lucky that the karts had equal performance and it was a very tight track. He would have kicked my ass on a real track for sure but hey, not today! Our luck was definitely in though and our teenage pro kart racer took us to an amazing victory, and I had the last laugh regarding the medium's prediction.

Whilst all the race teams took the winning ceremony as a bit of fun I wasn't. This was big for me and I was going to milk every second of it. I work for Jaguar and I'm in the Works F1 Kart Team on the top step accepting the trophy for 1$^{st}$ place. I so wanted the national anthem and a TV interview but instead we got a shed load of abuse and banter from the other teams. When I look back now it makes me smile that I was in the team that achieved Jaguar F1's only race win.

Post-race I was asked to stay for a meal and a charity auction and had some great fun. I started the bidding for a Chris Rea guitar and panicked when my opening bid paused for a few seconds but luckily that man Damon Hill was in on the action and put in a rather large bid. It saved my bacon as Danni would have killed me when she saw the credit card bill. There was also a raffle and tickets were going for twenty quid as I remember. A certain Vicky Butler Henderson was selling the tickets and when she came to the table all I had in my pocket was a tenner. I apologised and asked what I could get for my tenner. She took the money and planted a smacker on my lips. If I could have found a cashpoint machine it could have been a long night and even all these years later my wife still frowns at the TV when she is on!

With the car that started everything at the Reims GP Circuit.

Coming out of the first corners at Goodwood in my XKRS-GT.

My Jaguar sidekick Mr David Steele in the pits at Le Mans

A photo I took as we were heading onto the grid. A dream come true.

Pecos, Texas
High Speed Testing
with the XJ/SC

Alternative horse-
power in Texas.

Team spirit in the
Jersey Lilly Saloon.
From back left:
John Barnes, Me,
Richard Chant,
Paddy Healey and
Mark Wallace

168

The road to hell. The high speed track in Pecos

Getting into a facelift XJS, Dukes of Hazzard Style

A great moment in the Nevada desert doing 200mph with the amazing Michelle Rodriguez.

The gorgeous F-Type SVR cleaned and ready to be filmed.

Above: Being filmed in Phoenix.

Right: Scenic shots in Mexico on the way to Acapulco

Bottom: Camouflaged XF ready for me to drive on the ice lake

171

Out filming our new Jaguar XF with the legend that is Jay Leno.
Jono and I fly in to spend the day with Jay at his famous car garage.

Looking very cool at Pikes Peak. The F-Type FPP15 that got us 3rd in Time Attack. It was my proudest moment.
Even though we were keeping a low profile I managed to get a race number to match the project code for the car.

On the Mille Miglia with Richard, Bernhard and our beautiful Le Mans Racer, the Jaguar D-Type. The picture was taken as we waited for our holy blessing by the monastery.

The prototype build team with our Bonneville XFR

Anyway, back to Top Gear. The filming in Devon was a blast. I went down and teamed up with James May and also the late Mike Smith who was going to be doing the aerial filming in his helicopter.

James was amazing and so intelligent on a lot of subjects. We discussed the engineering behind the XF and then went into the theory of the Egyptian pyramids and various conspiracy theories whilst we enjoyed a morning cup of tea at the hotel. He was a thoroughly nice chap of the highest order and he made what was going to be a great experience even better.

My role for the event was technical support to make sure they didn't get or do anything wrong with our new car but when the fog came in and lost us a lot of time I ended up doing a lot of driving when James had to go back to London.

I ended up doing some great scenes that were used on the final edit of Season 10 Episode 10. It was an honour and when the episode was shown on TV the whole family came round to watch it with me. It was my first appearance on the screen, but it wouldn't be my last either.

On my return I did end up getting a ticking off for the cars registration. Before the event I rang the DVLA and sweet talked a lovely lady into getting me a specific registration as the car was going to be featured on Top Gear. They had VX57XFR and I did the deal. The only problem was that we had not announced that we were doing an XFR and PR were a tad miffed, until they saw the episode and then they were a bit more forgiving.

As a thank you for all the work in Devon I was invited along to the show at Dunsfold along with my then girlfriend (now my wife) Danielle. The episode we were on was the one with Lewis Hamilton and James Blunt and during the Hamilton interview Danni was standing next to him for the entire clip. It gave her a claim to fame moment that you can still see on YouTube and during one of our Goodwood weekends I took the liberty of taking Danni over to Lewis in the drivers' enclosure and introduced them to each other with the story about Top Gear. He thought it was really cool too!

Some years later after I had moved to the Special Ops division under the title of ETO or Engineered to Order I was working on the ballistic missile of the day the XFRS and we were fortunate to have been asked to support a film that was coming out where they wanted to use the new Jag.

As it happened, they were also going to film a spoof edition for Top Gear at the same time as part of the movie so that they could all use the same cars and save on costs. My boss at the time Kev Riches suggested that I go down to support the filming and act as technical support to assist the film crews for the film and Top Gear.

The location wasn't very glamourous though on the Isle of Sheppey but just being there was high up on the cool wall.

I met with both teams who were utilising an old industrial estate/oil refinement enclosure where the bad guys were going to fly out and make a run for it whilst being pursued by Ray Winston and Plan B. The runs were epic and the XFRS looked so good! I really wanted to be out there doing it myself and during all the hours of filming I kept on at the crew about letting me have a go and they finally cracked under pressure and allowed me to do one of the test runs. It was great fun made even better by the fact that the black Fiesta ST look alike chase car was being driven by a rally driving legend Mark Higgins.

With the film company moving onto the caravan park location in stepped Top Gear with that man again! His team were now going to repeat the scene from the movie for a complete spoof version to be aired later in the year.

They set up their own stunt drivers with Clarkson being the producer and they intentionally made a right pig's ear of it.

During the scene they decide to switch off all the traction control systems with the help of JAG MAN.

Clarkson stops his stunt driver, pulls up the bonnet and shouts for JAG MAN. I enter the scene and remove the appropriate fuse for my international film debut that was obviously worthy of an Oscar nomination at the very least.

176

We moved on to the caravan park scene for the final piece of the movie. In the actual film the XFRS and the Ford ST have a great chase and the stunt drivers did a great job on the grass to just keep the cars from hitting any of the static vans but the same can't be said for the spoof movie.

Clarkson decides he wants a monumental power slide from the Jag that should just clip the caravan before flying down between the two rows. At this point I must mention that the deal with the cars was that they can have light damage that can be fixed by a workshop but I'm not sure that he read that bit. We all stood just out of shot as the car came hurtling into view sideways and slid smack into the main structure of the towing hitch. It was a massive ouch moment and we all walked over to the car to take a look. It was certainly excessive and way beyond light damage for sure, but they didn't seem too worried at all as it gave them what they wanted.

I'm glad that I could go back to engineering and leave the discussions on the damage between PR and Top Gear.

During the filming for the final sequence I did manage to get some time with Clarkson, and we discussed all of our brief encounters and I wonder where we shall meet again one day?
I can feel a Grand Tour coming on!!!

# 16. Chauffeur to Sir Alex Ferguson No Less

As a result of holding the right permits and being trusted by our PR Team I have been asked to do a bit of chauffeuring now and again for a few well-known passengers. Sir Jackie Stewart was a regular when we were under Ford and I had to take him to a couple of drive events from his hotel as well as the current King of Jordan during a trip to MIRA, The King of Sweden and I also collected Prince Michael of Kent for a tour around Gaydon but I have to say as a Manchester United fan the biggest honour was being asked to chauffer for Sir Alex Ferguson.

I took a call from another one of my Jaguar mates and fellow footballer Ken McConomy who was the PR Manager at the time and he asked me if I could help him out with a driving job in London. My first thoughts were not great as London driving is not the best. He tried to secure the deal with an overnight stay at a hotel to save me driving back to Coventry overnight but again I was not over enthusiastic. Desperate to get me to do the gig he told me that I was on a Jaguar sponsored table at the League Managers Association Hall of Fame Dinner and I'm now warming to the occasion.

He eventually dropped his trump card and asked if collecting Sir Alex Ferguson and taking him to the event and then driving him to the station in the morning might secure the deal? Seriously, he should have just asked me that in the first place and reduced the cost of the call.
Ken was also a United fan and when asked about the driving job he thought that the best thing was to have a fellow United fan with a safe pair of hands behind the wheel and for that Ken I applaud your judgement!

I have been a long-standing United fan since 1971 when I first watched George Best play at Coventry with my dad. He was just the world's best player and seeing him score on that day was inspiring to a seven year old. Unfortunately my team had

178

quite a few ups and downs and we were missing out on the success that the club needed but then Alex arrived and the rest is history but not a lot of people know that Sir Alex's first game as a manager was against my home town of Nuneaton Borough who I have supported for nearly forty eight years. Alex was the new manager at East Stirlingshire in August 1974 after he finished playing at Ayr United. He took on the £40 a week part time job and put together a side for a pre-season friendly against the Borough and unfortunately for Sir Alex his new side came out second best.

Armed with this information from my Dad I was certainly going to ask him about it.

On the day of the dinner Tuesday 17th November 2009 I drove down to London Euston where I had been advised to collect my VIP passenger at a secret rendezvous point. The communications and logistics made it feel like I was in MI5 or a James Bond movie.

I was kept posted via texts with updates on his ETA and I in turn gave back my location and timing points.

I made sure I was in position with ten minutes to spare and pulled up the all new Long Wheelbase black Jaguar XJ in a space near to my collection point so that when I got my final text I could drive down and collect him with no fuss and hopefully avoid any press contact.

The text arrived to say he was walking to the collection point, so I drove slowly down the road until I saw him coming out of a side door and then I swooped in like the secret service collecting the President. With military precision I pulled up next to him as he appeared between parked cars and he hopped into the passenger seat and we made off as quickly as we could.

Well, there he was, as large as life sitting next to me and my heart was going like the clappers. He thanked me for meeting him and asked if I had a good journey down which was a nice touch. Our first discussion was how did I plan to get to our destination and what time would we get there? "Well" I said, "this XJ has the finest Sat Nav system already programmed in so it will take us to the front door and have us there in twenty minutes traffic depending".

He shook his head and advised me that we will not be going in through the front door and he would guide me around the back to a VIP entrance. How cool, I could only ever dream about this stuff.

With his mobile phone connected to the car he said he needed to make a call and tell them what time he was arriving. The line started ringing and then a Cockney voice came on and said "Oi, where are you?" Alex confirmed our current location and ETA and the familiar voice said he would meet us there. I had heard this voice many times but with worrying about the route and the traffic I could not place the name or face.

We had just started to discuss my role in Jaguar when his phone rang, it was his wife Cathy who was calling to discuss a pressing household issue regarding the rework of the central heating boiler. I guess being a Scotsman the quoted price made him cough and stutter a bit, but Mrs Ferguson was having none of it and advised him that it was getting fixed and that was the end of it. After the call he turned to me and said "See, I'm only human" and I laughed.

I was eager to discuss football but as we approached the hotel I figured it was better left until the morning.

Alex guided me around the block and then down a side street to the back of the Hilton, Park lane and we pulled up to some large doors. We both got out and I could hear the Cockney voice again coming through the door to greet us.

I opened the rear doors to get his bags out and then the voice appeared right next to us. It was Harry Redknapp the Spurs Manager. Alex introduced us and I shook his hand before he took the bags off me and said that he always plays bag man to Sir Alex. He then asked if I was joining them for a drink at the bar but I had to turn it down so I could get back to my own hotel and get ready for the evening dinner. They both thanked me again and said that they would see me later. What a moment eh?…I'm in the middle of two football legends rejecting the offer of a quick drink. So professional on my part, even in a once in a lifetime moment!

On arriving at my hotel and confirming to Ken that all went well we all got ready for the dinner and jumped into a taxi for the evening's footballing entertainment.

Every sponsored table had a Premiership Manager on it and for the evening we had another larger than life legend in Sam Alladyce who kept us entertained all night with some funny stories and made sure that the wine waiter didn't miss us out.
On the table next to us was David Moyes, the Everton manager who went on to replace Sir Alex at United, who was getting quite a bit of flak and banter from Sam.

It was a special evening as Sir Alex was the guest speaker along with Harry Redknapp and I sat glued to his every word as they talked about football life. He didn't mention Nuneaton Borough though…. but I would get him on that one in the morning.

Later on in the evening I had a tap on the shoulder, and I noticed that all the jaws dropped opposite me. I turned around and Sir Alex said "David, are we ok for an early start tomorrow?" Well, I was here to serve so yes it was absolutely fine by me and he confirmed a 6am meeting in the reception shook my hand and retired for the evening.
Feeling and looking like a Cheshire cat I also said my goodnights to the table as duty calls and I must get some sleep.

When I think back to that moment it makes me laugh. The Godfather of footballing greatness not only stopped at our table to discuss travel logistics he remembered my name!! Thank you, Ken and Jaguar, I love you both for that moment.

As it was so early, I parked at the front of the hotel to collect him and the doorman had no problems with me leaving the car at the door until he was ready to leave which was extremely handy.

Alex breezed into reception and we quickly made our way to the car and took off for the station.

I was puzzled though because he came out of the lift from his room, said goodbye to the receptionists and just walked out without paying the bill. I had to question this before we drove too far and he just laughed and said, "Don't worry, it gets taken care off son." How swanky is that?

He seemed to be very relaxed and was asking about the car and Jaguar and my role at the company, so I thought this is it...my chance to ask some questions. I first confirmed my allegiance to the United badge, as you do, and then dropped in the question about his first ever game as a manager and could he recall the team he played.

I was surprised that he remembered the dates and that it was Nuneaton Borough and he also told me he kept the programme. With raised eyebrows I asked about the score line to which he turned and glared at me "aye, we lost".

I told him all about my family connections with the club that I had supported from when I was 7 and that I was helping out doing youth team coaching at the club to get my coaching badge. He spoke fondly about the Borough and had obviously kept track of them over the years which impressed me.

I was prepared for our encounter and had photocopied the local newspaper article about that game which he signed for me with "Up the Boro ...Alex Ferguson". He also signed one of the dinner invites to my close United fan Martin who was the most envious man on the planet that day!!

I dropped him back off at the secret side door and shook his hand wishing him and the team every success for the next season and very cheekily said that next time I was up at Old Trafford I would knock his door for VIP Tickets.

"Aye, you do that."

What can I say, it was just one of those very special one off moments that you treasure forever. I still regale to anybody who will listen about my meeting with the great man.

It was an honour and a pleasure of the highest order and just in case you are ever reading my book Sir Alex, I never collected any VIP tickets by the way!! Anything is still possible you know!

To Ken, you are as big a United fan as I am and to pass on the gig to me says a lot about you big fella.
You are a Jaguar Legend mate.

# 17. Creating a GT

In my time at Jaguar I have been asked to do all sorts of interesting stuff, but being asked to redesign a car was something else. It all started with a call from my old mate Steeley, who was now at the dizzy heights of corporate management, asking me if I could pop over to discuss a project that the chairman was interested in.

I went over and sat with his team and the problem they had was that the currently released XKRS was getting great reviews but the press still didn't rate it well on the track, so the chairman was keen to explore what we could do to the car to improve its credibility. Steeley knew that I had just joined the Special Vehicle Operations team and handed it over to me telling the chairman he was confident we could do something. No pressure then!

I was really excited about doing this and I knew that I would have to put a project together outlining our intentions to my senior management for approval. The best way to start though had to be with a design image of the final car.

Over the years I had made some great contacts within Design including the legends that are Ian Callum and Wayne Burgess, who were both petrol heads and loved getting involved with skunk projects. When I mentioned it to them they were really keen and allowed me to continue the work with a guy called Joe Buck. Like most designers, Joe was pretty amazing when it came to design sketches and we also had some history with other skunk work including the painting of a Jaguar plane.

I outlined our plan to turn the XKRS into a track car by focusing on vehicle dynamics and aerodynamics to produce more grip and downforce. We decided that we would need a better front spoiler, some aero aids and a bigger rear wing. With the base idea in his head, Joe produced some great images to take forward to the management team when we were ready with the rest of the data.

So, what do you do next? Well, you have to call in some favours with all your mates and contacts in the business, because this is most definitely not a one man band!

The biggest part of the project had to be aerodynamics and we had a great guy called Javier Castane. He was from Spain and was a passionate Barcelona fan, but we never held that against him, although we did have fun with his accent as he sounded just like Manuel off Fawlty Towers. I remember asking him if he fancied a coffee and a chat and he joined me in the canteen where we went through the outline of my plan. He was well up for it, even though most of the initial work would be outside of the core business in true skunk style. We went through the XKRS base line numbers and Javier advised me that no car in Jaguar's production history had achieved positive downforce. Even the legendary XJ220 produced zero lift and zero downforce, so our task with the XK was to both reduce lift and generate downforce.

We discussed front splitters, rear diffusers, rear wings and air dams but I was also keen to review the bonnet apertures, making it easier for air and heat to escape from under the bonnet. I also wanted to try something else that I had seen, not on a car but on the Typhoon fighter plane with its front nose canards. The pilots were impressed with how they improved the front-end manoeuvrability of the aircraft and I asked Javier if we could do something similar to create some downforce. He already had loads of ideas and went off to play!

The next step in the dance had to be chassis dynamics and I was fortunate enough to have some close friendships with the guys who had completed the XKR and the XKRS, and were now already working on its replacement, the F-Type. These guys were not cheap and it cost me a lot more on coffees and a walk over to the main site, but it was well worth it. The team would include our dynamics specialist, David Pook along with his wingmen, Chris Spindler, Dave Hudson and Mr Dampers, Matt O'Hara.

I explained that we had to do something really special because we did not have the option to increase the horsepower of the engine.

185

My brief to the team immediately sent Pooky off the Richter scale and he started spewing out all sorts of ideas that sent me into a flat spin.

With four large Costa coffees downed they all agreed to join the skunkworks.

Putting the team together was just like something out of a movie, along the lines of Mission Impossible, The Magnificent Seven or The Great Escape, although I'm not sure the endings of two of these went too well.

What I needed next was somebody to do the interior trim. With the additional grip and improved aerodynamic package we were going to need better seats, steering wheel and some extra finishing touches to compliment the final proposal. I approached another close mate from my time on XKRS, who was also a fine petrol head and shared my passion for military aircraft, by the name of Mark Twomey. Mark was from Ireland and was a very calm and relaxing person to have involved. He loved Special Ops and when I told him about the skunk project he was in. I didn't need to buy him a coffee either!

My last team member had to have experience doing body panels and exterior ornamentation and my sights were set on a chap called Mark Evans. Mark had a wealth of experience having worked on the XJ220, so was the perfect candidate for a canteen coffee chat. As it happened he didn't drink coffee so we had a nice cup of tea to seal the deal. Mark also suggested a guy that I didn't know too well by the name of Dave Foster who was a bit of a whiz on Computer Aided Design which would be perfect.

On my drive home that night the theme tune to the Magnificent Seven was playing in my head and on the current count that's how many team players I had, but I knew we would probably need a few more.

As I sat at home hatching my plan I realised I would need a programme office person, who could control the timing plans and generally knit the project together to take it into a limited production run, and there wasn't a finer person to do this than

Pete Page. He was like a brother from a different mother and that was clearly evident by his ginger locks!

I called him at home to discuss the project and he loved the idea as much, if not more than the others. Pete had a great working relationship with the build teams at Solihull which would help us out a lot if we could get the concept to production. He also had a wealth of knowledge on finance and releasing parts into the system. He would end up being my biggest and highest paid signing for the team as he loved malt whiskey and we ended up drinking quite a bit of the stuff during the project (after hours of course).

With the initial team in place we held a lot of secret meetings and came up with all sorts of ideas. When we agreed on our specification for the car I completed a full-blown presentation to run past my boss, Kev Riches and our Director, Pete Simpkins. It was received well and I was given the thumbs up to produce a prototype and create a plan to produce a limited edition run. With the backing of my direct management I presented our ideas to the Chairman and Steeley, who also loved it.

Having been given the go ahead we were now able to move from afterhours skunk works to a full-blown project. The team were thrilled by the news and they in turn threw themselves at the project with a passion.

Javier moved into the wind tunnel at MIRA armed with all sorts of things to try. He had cardboard cut outs, bolt on meccano and bits of string to attach all over the car trying to measure optimum air flow. At the end of every session the team gathered round to go through the data and in no time at all we were building an aerodynamic prototype with more robust parts that were showing great signs of improvement with each individual test.

The front splitter was adding a lot of downforce as were the new air dams that were in front and at the rear of the wheel arch. My Typhoon canards where adding downforce and were also having an impact in directing air through the front gills on the bumper. The front-end data was now causing us to review the rear end and the team were calculating just how big the rear wing had to be. Mark Evans suggested that we pass this info to Dave Foster, but he was busy on other projects and the only thing he could do was take it home and complete it after hours.

He spent many hours locked away at home and his final design that curved to match the rear of the car was just gorgeous.
We mocked up the wing and returned to the tunnel and the results were great. With the modifications completed so far we had converted a vehicle with front and rear lift into a proper track car with actual downforce.

Our next task on the aerodynamic list had to be the bonnet. We turned our attention to the C and D Types from the 1950's and then the E-Type from the 60's. It was quite clear to see that pressed louvres were the way to go rather than the stylish vents of the modern cars. Mark Evans took on the task and worked with our suppliers to press us a prototype bonnet and the rewards were instant. With two rows of slats, the air going in through the grille could now escape easily improving downforce even further. Prior to the modification the bonnet was acting like a sail and when we did the actual testing the cooling performance improved as the hot air could get out much faster.

Down in Chassis Development we had been using an XKRS to tune the suspension. Pooky and Spindles had been really busy and come up with an Eibach Motorsport twin spring system. The front spring rates were up 68 percent and the rears up by twenty five percent. Matt had also been busy to get us a height adjustable bespoke adaptive damper system, and the base car was transformed.
During our meetings we were also aware that we needed to reduce the corner weights and the question was raised about using carbon brakes from one of our other development programmes, the new F-Type.
The core F-Type team were in the early stages of development and after a lot of pushing from our side they eventually saw the benefits of the GT taking the system into market first, and the car would really benefit from it.
With the bigger disc brakes we advised Joe Buck that we needed a new wheel, which he was keen to do from the start,

and he came up with a very aggressive forged five spoke 20" wheel just for this car.

Now that we had the suspension, brakes and wheels, we had to go knocking on the door of the Traction Control team and call in some favours from our spin doctors, Andy Gosling and James Rawlinson. They had played a major part in traction and stability control for many years and were also a very useful set of hands behind the wheel. Like all the others they were more than keen to get stuck in from start. Using the chassis development car they went over to the Nurburgring with Pooky and started the process of black magic. It's all to do with binary digit coding into the various engine management systems and it really is a black art!

It was great having them on-board but my Magnificent Seven were now developing into the Dirty Dozen with the addition of another head to look after the carbon brakes.

With plenty of dynamic testing going on, Mark Twomey had been making some great progress with the bucket seats and restraint system, and had even started to look at a limited version just for the UK with half a roll cage. We were definitely getting close to producing our one and only prototype.

Working alongside Pete Page we had managed to get other little GT tweaks into the programme with unique badges, bonnet graphics, carbon door mirrors, blacked out bumper bar, carbon rear diffuser, modified active exhaust tune and an interior carbon graphic set. Pete worked so hard to get all the parts released into the system and also chaired the team meetings that kept us all on track whilst I focused on the engineering sign off and launch planning.

It was a lot of dedicated hard work, but we were also having so much fun doing it and that makes the world of difference. For me I couldn't wait to drive into work and crack on with the day's schedule and the whole team had a smile on their face. We were also getting a lot of attention from all the other projects in JLR and lots of questions were being asked about what we were doing, but we loved every second of it because it just makes you feel a little bit special.

With the car built it was time to have a drive on the track and it didn't let us down at all. I remember going out and doing several laps at Gaydon getting faster and faster and pushing it around the bends much harder than previous vehicles. It was just awesome and the level of grip from the Pirelli Corsa tyres changed the whole dynamic of the car.

This was a very different Jaguar that made you feel more like a fighter pilot. It was precise and accurate around the steering with impressive body control. The stability control systems worked really well and the whole experience was leagues above the base XKRS. Pooky and his team spent hours and hours at the Nurburgring to get the car where it was, and it showed in abundance. This car had an edge about it that made it feel alive.

With just about all our testing complete it was time to send it out to the Nurburgring for the final tests and see what sort of lap time it was capable of.

With a lot of help from Phil Talboys, who was our resident engineer at the ring, the car was shipped out and the dynamics team went out to start doing some very fast set up laps before we were to hand it over to our very experienced Nurburgring race drivers, Dirk Schoysman and Vince Radermecker.
Between them both and with some help from Pooky the lap times started to come down as they gained more and more confidence with the car. They started at eight minute laps and it wasn't long before they had brought it down to seven minutes and fifty seconds. The laps kept building up and the stock of tyres were going down as was the lap time. When the news came through that we had completed a seven minute, forty five second lap we were over the moon, but the team were desperate to push it further. After a lot of blood, sweat and tears, the boys managed to get an epic lap time of seven minutes and forty seconds, knocking off eleven seconds from the XKRS lap time and also making this car the fastest ever

Jaguar around the circuit, beating the previous record of seven minutes and forty six seconds by the XJ220.

We were jumping for joy at the news and so thrilled to have achieved so much with just a handful of dedicated engineers. All the top race teams would spend millions in development to find tenths of seconds, yet here we were with a very limited budget and limited resource making massive improvements to our lap time.

To go with our new record the car also managed a 0-60mph time of 3.9 seconds and had an electronically limited top speed of 186mph.

When we announced the news to the team it was just like we had won the World Cup. All the dedication and attention to detail had paid off and we were all on cloud nine.

I was getting emails from all over the place congratulating us on our achievement and I was incredibly proud of the whole team.

All that was left to do now was for myself and Mr Page to finish a very expensive bottle of malt whiskey and work out how we were going to make a batch of production cars.

And now the fun starts with a request from PR to go to Millbrook Proving Ground and do the filming for a launch video. I head off for a whole day of filming on the alpine circuit and the high-speed bowl. The results were fantastic that can be seen on good old YouTube if you type in 'The New Jaguar XKR-S GT.'
I still watch it every now and again just for the smiles.

After all the data had been checked and signed off, I was given the task, or should I say the privilege to drive the car at the Goodwood Festival of Speed. It was an honour to be sat at the start line in something that you have created with thousands of eyes and cameras ready to catch their first glimpse of Jaguar's fastest road car.

As part of the weekend I would do five of the six drives because we had a guest driver for the other one that was none other than Sir Chris Hoy. He was just getting into racing and was chuffed to bits to be getting a go in our car. I spent quite a while with him going through the course and warning him about the very difficult Molecomb corner before taking him through the details of the car. He did a good run and brought it back safely but the following year he must have forgotten all my words about Molecomb as he went straight on into the hay bales with his Nissan GTR Race Car.

The car was subsequently used in America at Willow Springs where top race driver Randy Porbst did a flying lap and was very complimentary about our achievements, as was our own Chris Harris of EVO and Autocar fame who, on a damp Oulton Park Race Circuit, said that the car was sensational and that the Jaguar skunk works era had begun.

The car also managed a bit of celebrity status as a guest at RAF Scampton with the Red Arrows. Obviously I had to accompany the GT for this and ended up doing a few demo runs up the runway and a lot of photographic work. I may have said it before, it's a dirty job but somebody has to do it!

After all of its hard work and as a thank you to the teams that delivered all the carbon parts I signed off the car to go to the suppliers and be put on show in their main reception, but it never arrived. My body development man in the shape of Mark Evans was driving the car on the M40 in very wet weather when he hit a lot of water and ended up having an altercation with the armco barrier. He was mortified and it took us a long time to convince him it wasn't his fault.

His phone call to me that morning from the motorway is now part of Jaguar history and very funny! I answered the phone and poor old Mark was in a flap saying he had damaged the car. I questioned the damage and he replied that he had hit the barrier. Within Gaydon we have security controlled barriers so I asked which one he had hit. He replied that he had hit a few barriers that left me scratching my head! I could not understand how he could have hit multiple security barriers. I told him to stay put and I would walk down to take a look,

Noooooooooooooooooo was all I got back at first followed by "The Police have had to close the road to get the car moved." Eventually we got there and the information confirmed his attraction to a great length of the M40 armco barrier.

The car was recovered, re-built, and is now back with me in the Classic Car division enjoying a rest but still getting lots of attention with every tour to see the collection.

Mark also joined me at Classics and we are back working again on some Classic Skunk Projects this time.

I have to say that it was an absolute pleasure to have delivered this car and my thanks go out to all those who helped me and the team to make something a little bit special, but I would like to finish this chapter with the words from our brilliant design director, Ian Callum, who captured what we had achieved.

"The XKRS-GT has been designed purely by the laws of physics. It has been developed in the wind tunnel and on the racetrack with the sole aim of creating as much high speed stability and rear end downforce as possible. Nothing has been styled for the sake of it. It's been an exercise in efficiency and the result is a car that's raw, focused and devastatingly quick."

Great words from a great man about a great car!

So, what's it like? Well, I got to drive the GT a lot and it just never failed to deliver on all fronts. I did some great drives that included a trip into Wales with EVO magazine, some cool runs up the hill at Goodwood, a video shoot at Millbrook, track days at Oulton Park, Castle Coombe and Silverstone, a day out at Scampton with the Red Arrows and loads of laps at Gaydon but the one drive that stands out was a blast out on the roads around Kineton near our Gaydon HQ.

My boss at the time, Kev Riches, had mentioned that our purchasing manager had never been in one of our cars and maybe now was a good time to get her into the GT. I gave her a call to arrange a drive out and even on the phone you could tell that she was thrilled to bits.

I have to say that drives like this are the best because you really get to gauge the reaction from people who never get the chance to drive super-fast cars.

I took the car around to the front of the building and there was Charlotte, smiling like a Cheshire cat.

I pulled up and hopped out to give her a quick tour around the car and show her all the things that her and her team had paid for. As we worked our way around she couldn't help but be impressed and kept telling me how excited she was to be having a drive out in it.

I helped her into the passenger seat and her first comment was "What the heck am I supposed to do with all these buckles?" The GT was fresh back from the Nurburgring and was still wearing its five point racing harness system. I looked at the harness then looked at Charlotte and returned to the harness before saying "I know we hardly know each other but if you need me to sort this harness out then it's going to get a little bit intimate."
I brought the top straps down to the centre buckle and she had already started giggling.
The side straps were next and thinking that was it all I heard was "Crikey, well now I'm well and truly strapped in!" I lifted up the centre crotch strap and motioned as to where it had to go which set her off into fits of giggles again.
I strapped myself in by which time Charlotte was just about in control. I pressed the start button and I don't know who made the most noise? The car with its tweaked exhaust system was truly epic but the shriek from my passenger was just as loud! "You don't get out much do you Charlotte?"

As we rolled along to the security hut at walking pace Charlotte was already singing the praises of the exhaust system and just how loud it was. I started to smirk because I knew what was coming when I give it the full beans!
We made our way out onto the main road and making sure we were clear of the site entrance and free of traffic I unleashed the GT. Normally I would be enthusing about the colossal amount of exhaust noise and the performance feel of being planted into the back of the seat as the track tyres try to rip off the top surface of the road, but not today. My passenger had other ideas and went into full over excited mode! As the car leapt off

the line there was a massive shriek followed by fits of giggles and then a mixture of nervousness and adrenaline fuelled emotion.

Flicking through the gears using the paddle shifts on the steering wheel we start to eat up the road at a serious rate of knots to hit the speed limit in about 3.5 seconds, but I think Charlotte was still delayed by a couple of seconds. When she did catch up she couldn't stop laughing.

We took the car around the local B roads for about twenty minutes and I can honestly say that Charlotte shrieked, laughed and giggled the entire way round.

We pulled up back at reception and I helped her out of the car. She was still shaking with excitement and could not stop thanking me for taking her out. She was raving about the car, the speed, the noise, the speed and then more about the noise, the safety harness, and the speed again and then even just a little bit more about the noise.

In hindsight it's a good job I didn't take her out on the test track and restricted the occasion to just an enthusiastic drive out in the countryside.

Well her reaction said it all really and that makes you feel great.

Some six years later I bumped into Charlotte who smiled when she saw me and her first comment was that she was still giggling about her go in the GT.

Job done I reckon!!!!!

# 18. Fast and Furious in Nevada

There have been lots of great things and good reasons for having brilliant memories with Jaguar. The fabulous people and work colleagues, amazing cars, stunning locations, top celebrities and epic testing form the basis of each memory and individually contribute to various chapters in the book but what happens when you put all of these together for just one uber cool moment in time? Well, it kinda goes like this.........

Its early 2016 and I'm at the end of one of our finest SVO projects, the F-Type SVR. We have completed all the testing and are now in a period that we call the Launch Phase. This is when the pre-production cars go out with the PR and events teams to create media attention and engage with the press. Leading up to the launch there was a lot of attention around the top speed of this car being 200mph and it was decided that we should create a launch video that shows the car doing exactly that. During testing we had achieved the top speed at the Nardo Test Track and we even pulled an unofficial indicated 216mph on a very fast downhill section of the German Autobahn when the dynamics team had finished Nurburgring testing. Take a bow Spindles and Hudson, or should I say "Read the Badge!"

Getting to 200mph relied on having a long enough bit of tarmac, as the car raced up to 190mph quite quickly but then the aerodynamics slowed it down a little and the final 10mph took a bit longer.

Following all our testing we held discussions with Sales, Marketing and PR and concluded that rather than letting the press out on track or road trying to get to the top speed we should engineer a test and film it as evidence that they could keep. The idea was well received and within a couple of weeks my boss at the time (Andy Smith) approached me with the most terrible news that he was going to have to send me out to Nevada to co-ordinate and drive the SVR at 200mph. He was gutted that he couldn't do it and he could barely move his lips when PR asked him to tell me who I was going to be working with. Not phased in the slightest I hippidy-hopped back to my desk to the sound of 'I feel good, durrr de durdle lerrrr, you know that I should durrr de durdle lerrrr' by the one and only James Brown.

I was to fly out to Vegas where I would be met by one of the production team who would then take me out to a spot in the desert not far from the infamous Area 51, on a stretch of road called the 318. My job was to then take delivery of our new rocket ship, the F-Type SVR, and prepare it to run on route 318 at 200mph. When I was happy that the car was fine, I would then take our Hollywood Superstar out for a drive to make sure they knew all about the car and felt comfortable driving it. With that completed I would handover to a third-party race driver/instructor who would passenger and coach the film star to 200mph whilst I joined the film crew to drive the chase car with the cameraman and get some great footage.

Armed with all this information I now had to go home and convince my wife Danielle that this was a fantastic opportunity that I couldn't turn down and that I'm sure that she, the two kids, two dogs and six fish would all be fine! I needn't have worried really because over the years she is used to it and very understanding and incredibly supportive of all my crazy adventures. She has not done too badly out of it either as Jaguar has also been very kind and encouraged her to attend some of the events with me. She adores the Goodwood Festival of Speed where I did manage to take her for a race up the hill when I had a spare seat. That drive was also special because she held on and was very quiet until we parked up at the top and she looked over and said "oh, so you do know what you are doing then". It was a little bit of acknowledgement that it wasn't all just fun and games.

I managed to introduce her to a number of celebs at Goodwood including, Jay Leno, John McGuinness, Jess Ennis-Hill, Lewis Hamilton and Sir Chris Hoy and she enjoyed her brief time with each one although I think her own highlights were bumping into Matt Dawson the England Rugby star and her ultimate idol Kevin McCloud from Grand Designs.

She also enjoyed the Goodwood Revival dressed as Amelia Earhart the American Aviator and being asked to pose next to an old biplane as they said she looked perfect (she always does).

So, I set off from Heathrow on a cool March morning heading out to Vegas all on my own as the film & support crew left the previous day. The flight allowed me some time to sit and read up on our movie star. I always like to know a bit about them so that you can have better conversations when you meet them and there was quite a bit to read on this one so it should be fun.

In the days before I left I was given a contact that would meet me at the airport and she had already texted me to say that she was already in Vegas collecting provisions whilst the rest of the team were out on location working out where to film. All I had to do was head to the car park where she would have a car ready to drive us both out to the desert.

You have to admire just how professional these teams are, and our PR and support teams are the best by a country mile.

The attention to detail is at a different level as preparation is everything to a great event and as I was once told by a very wise RAF commander PPP = PPP or Piss Poor Preparation = Piss Poor Performance, and he wasn't wrong.

I arrived at a very warm Vegas Airport and followed the detailed instructions of where to go and found myself at the correct level of the car park faced with two females, both with cars waiting for collections. On the right was a lady who looked like she could take on Mike Tyson for a few rounds and on the left was a very attractive lady who looked every inch like a film star. There were a couple of seconds of tension that felt like minutes as I waited for some verbal contact to confirm my fate and boy what a relief when the potential film star said "Hi Dave I'm Britt, are you ready to go?" Phew what a relief.

We left the airport and headed out onto the freeway and within ten minutes we were gassing away like we had been friends for years. Danielle is always saying that I will talk to anybody and she is right. As a kid I used to be really shy and would not talk to anybody but as I grew older and gained more confidence, I found it far more comfortable starting and holding conversation. It's a lost art now at work, as many of the new graduates are brought up with social media and email and all their conversations are electronic. When I first joined Jaguar, I never called somebody if I was close enough to walk and see them face to face.

Britt and I chatted about all sorts of stuff and the conversation turned to the event and our destination near Area 51, which lead me into one of my many interests on the subject of UFO's and conspiracy theories. We exchanged our own thoughts on both subjects and Britt found it fascinating, so much so that she missed the exit off Interstate 15, and in America that's not great, as you end up driving miles to the next junction. We decided to continue and then get off and go cross country back to our destination. It added an hour to the journey but that gave us extra chat time to use wisely. Funnily enough one of the things we discussed was that I had started to write a book and I was struggling to get time to write it. Britt was full of encouragement and passed on a lot of great advice having done some herself and here I am now some three years later actually finding the time to get pen to paper.

By the time we arrived at our swanky desert hotel it was very dark and the short walk from the car park left me speechless as I stood and gazed at the sky. I have never seen quite so many stars. The location was perfect as there was no light pollution for miles and miles and with a warm breeze and the sound of desert critters it took me back to being in Texas. There must be something about the open spaces in the desert that just appeals to me, maybe I was a cowboy in a previous life? Although it was late the team were good enough to stay up to see us in and have a beer. There was Mary Zeeble our American production co-ordinator who found the location and set it all up, Louise Allen our head of production, Ben Seyfried the race car driver and instructor from the UK, Al Clark the cameraman, Tim the director and then Andrew Thomas and Nikki Johal all from FP Creative who will put the whole show

together. We were also being joined by one of my favourite PR mates Juliet Fairbairn to oversee the operation.
It was good to see everybody but after a beer or two it was time to hit the sack.

The next day we set off to route 318 and made our way to our base location of a working ranch where Mary had set us up with a barn to work out of. It was absolutely perfect but unfortunately the real star of the show had not arrived. Our F-Type SVR was still on the back of a truck and was nowhere to be seen. After a quick team meeting the film crew and I decided to go out on the route with a very nice Mustang and work out what we were going to do and take a look at the stretch of the desert for our high-speed run. The actual ten miles of road that we were going to use is part of the 318 that is also used once a year for an open car race that Guy Martin went on to do in 2016 for his Channel Four TV programme. It's fairly straight for as far as the eye can see but does have the odd bump and camber to keep an eye on. We end up finishing our recce and head back to the ranch for an update only to find out that the truck got lost and will not be with us until early morning which did add a bit of pressure but there is no point in making a drama out of it. We all decide to carry on with the set up so that our only focus in the morning is getting the car up and running.

When we are done its back to the hotel via a quick detour onto the Extra-Terrestrial Highway to take some pics and go all X-Files!
I give the team the full run down on Area 51 and they are hooked. It's a shame we are tight on time as we could have gone for a close visit and a bit of top secret snooping via the Little A'le'inn at Rachel. This place enjoys a bit of celebrity status having appeared on many UFO and Area 51 documentaries and was even featured on the film Paul.
If only we had more time! I could so write a book about this stuff!

Back at the hotel I meet another member of the team called Donnie who has been sent by Jaguar North America to help us with vehicle movements and logistics. We both decide to hit the sack early so that we can do a 4am start and get out to the ranch to meet the truck driver with our car, hopefully! With a good night's sleep, we slip out of the hotel as planned and head off to the ranch only to find that the truck still hasn't arrived and Don is straight onto the trucking company to get the drivers details and phone number. It turns out he wasn't far away and was still sleeping after a long drive the day before.

When he arrived we expected a car transporter and an easy unload, but no...that would be too easy wouldn't it? The SVR was still in its wooden shipping crate that had been dropped onto the truck by crane. Don and I looked at each other shaking our heads and then both looked to the far end of the ranch at a very large forklift, which might just do the job. Without further ado it was time to wake the owners of the ranch and although it was still only 6am they were more than happy to help and even started laughing when I mentioned the old forklift, which obviously hadn't moved for a while. With a bit of encouragement, some fuel and a new battery it fired up and we were away.
The forks were slotted under the crate and we carefully lifted it up realising that the box and car were a bit heavier than the forklift. With some extra bodies on the back of the forklift we raised it just enough for the truck to move out of the way so we could very slowly lower it to the ground.

Our next challenge was how to open the box but our hosts came to the rescue once again as their son, who was a part time fire fighter, had all the kit we needed in his barn and it wasn't long before he had cut us through the box and released the caged cat!
By the time the team arrived she was out and on show in the gorgeous morning sunrise and Don and I were ready for the first brew of the day and some very welcome sausage and bacon rolls.
By now the whole family were out and staring at the new Jaguar and they were all very impressed by its looks, but the best was yet to come. I got in and pushed the button to fire up the V8. Once it was warmed up, I gave it an enthusiastic tickle on the accelerator and the crackle from the exhaust was epic. A quote

from one of the ranchers was "God damn that is a mighty fine noise you got there". I have to say it is a very special noise and we love it.

The rancher's wife was telling me she had a Corvette and it was probably faster than the Jag and maybe sounded louder too so, as I was itching to get out for a drive, I suggested that she comes along with me.

We head out from the dusty ranch road and onto the 318 with not a soul in sight for miles so it would have been rude not to give it the beans. The SVR burst into life on demand and in next to no time we were thundering down the highway and as I approached a suitable turn point I used the paddle shifts to change down through the gears causing a wonderful exhaust crackle and pop with every change. What can I say, I'm an adolescent child at heart and I love it!

With even more beans unleashed for the return run we are soon over 100mph and then downshifting again to create more noise for the crowd at the ranch.

I pulled up and the grin from my passenger said it all…Jaguar 1 Corvette 0 and that takes care of public relations for the day.

We spend the next hour checking over the car and putting on some new tyres as the ones fitted are just used for transporting the car on the plane and the trucks. All fluids are checked, tyre pressures double checked and I'm ready for the serious stuff to start and I head out into the vast expanse of the Nevada desert to meet Mary and the team who have by now got the Department of Transport support crew there who have closed nearly 10 miles of the desert for me.

This has to be one of the coolest experiences having ten miles of desert all to yourself. No cops, no side roads, no public and fingers crossed no issues. On the previous day I selected the most suitable stretch of road available to us making sure it had the best bit of tarmac, no other ranches to go past and no ditches at the side of the road that would cause you a whole world of pain if you went off. You just couldn't do this in the UK.

We had approx. 90 miles of straight road to choose our spot, it's just unreal.

With only one truck going through the transport guy at the other end confirmed it had passed him and the road was officially closed with the aid of two large lollipop stop signs and two Department of Transport guys wearing the most amount of fluorescent gear you have ever seen.

For the first run out I decided to take it up to 170mph to see how it held on the road and where the best position was. Either side of the centre white line there was a slight camber, so the best position was straddled over the line.
The run was a breeze and the car felt awesome, so I pulled up to get ready for the return. I looked back south down the road towards the mountains and I contemplate for a minute or so about the return run. I never think about what might or might not happen to me when I do this sort of stuff, but I do think about all the things that may affect the run. I look out across the plains checking for cattle and then look for dirt devils that are like mini tornadoes that might sweep in front of you. I check all the gauges in the car and then confirm over the radio that all is clear and I'm ready to go again. I take a deep breath, relax my shoulders and give it some more beans. I pass 170mph quite quickly but the next part of the speedo from 170mph to 200mph takes a little bit longer as the soft air that we breathe suddenly becomes an invisible brick wall trying its best to stop you in your tracks. The SVR is also doing its best to eat as much of the air as it can as the supercharger is scavenging everything that comes in through the grille. The rest of the air either flows over the cars gorgeous lines or forces itself under the car in an attempt to cause lift and make you take off like a jet fighter. If all the engineers and aerodynamicists have got the sums right the car should stay firmly in contact with the ground.

I'm now eating up the tarmac at a rate of nearly 300ft or 91m per second. It's a massive distance when you think about it. In the time Usain Bolt would have run 100m the car would have travelled nine times that distance. It's insanely quick and my adrenaline is flowing just as fast now and it's at this point that you tend to start gripping the steering wheel too tight and you have to consciously relax your grip and drop your shoulders a little and then have a quick look at the speedo which is now at the magic number of 200mph. I start to lift off and let the car slow down nice and easy with no braking and eventually I'm at a very sedate 30mph and I can now take my next breath.

As I roll up to the team at the stop sign there are a couple of trucks and few cars waiting to use the road which has been closed for about fifteen minutes but that doesn't stop them from honking their horns and yelling some abuse at us all as the GO sign is shown to them. I guess we had better call that a day then as we don't want to cause a major incident and force the desert road into chaos or gridlock. Not sure what they would make of the M25 at rush hour, but they sure have it easy out here!

With the car back on the ranch we let it cool down a bit before completing a ramp check to make sure all is ok.

Donnie does a very thorough check and reports back to us that all is good for the runs tomorrow so all we have to do now is clean up, put the car to bed and enjoy a gorgeous sunset with a cold beer, some great food and just the best company ever.

So it's the day when our movie star arrives and although I knew who it was before I left the UK, I figured it would be good to keep it secret for the book and build up the suspense a bit.

Arriving at the ranch the team have already sorted out breakfast and coffee and there is a buzz around the place. Al Clark has been busy setting up the Ford Mustang as a chase car with an amazing camera rig on it which I will drive when we start filming. Ben has also had a few runs in the car to make sure he is happy. Trying to get him out is hard though so he must like it. We are all discussing the plans when the noise of a helicopter interrupts us and makes its way to land on the ranch. It's a great looking thing with a massive camera on the front that costs a fortune. The film crew have arranged for the aerial support and they look very professional. When I go over to take a look Al tells me that they have just done some filming with Ken Block in his WRC Rally car and they are the best in the business.

The helicopter crew join us for a debrief and we work out all the plans for the days filming. I'm so impressed with all the detail and thought that is going into the video. I would have liked to have done the 200mph run with our VIP but unfortunately the liability of a JLR person in the car, should anything go wrong, is very high, therefore a third party race driver and instructor is far safer, and I understand that.

It's not long before we hear the second helicopter that has our guest on board. It pulls up around the ranch and lands nicely next the other helicopter and we wait for the engines to wind down and the blades to slow up a bit before walking up to meet our movie star guest. The door slides open and there she is looking so cool in her leather jacket and shades. It's none other than the Queen of the Fast and Furious films Michelle Rodriguez.

We do some quick introductions and she heads straight for the Winnebago van to freshen up but it's not long before she returns and hops into the passenger seat of the car so that we can drive out to the road and do some familiarisation of the location and talk about the car.

Michelle was very relaxed which was a good sign and she started asking me about who I was and why was I out here with the car and it was good just chatting about normal stuff. As we drove north on 318 I started to discuss some of the important things about the car appropriate to what we were doing really. As I started to tell her about the aerodynamics, she interrupted me to ask about rear lift and were there any issues? A good question because most cars have lift and I told her it was more about balance and there were no issues with this car.

The next set of questions were around gearshift strategy and how the gearbox decided which gear was best to use. This was followed with questions on the paddle shift system and then tyre grip. They were all good questions for what she was going to be doing and I was impressed with her knowledge. This continued for the next ten minutes and I must admit that I found it quite intoxicating from an engineering point of view (honest). There is nothing better than being well educated, well informed and articulate on your chosen subject. I have been in and had similar conversations with Jessica Ennis, Carol Vordeman and Rachel Riley who all impressed me with their level of engineering knowledge and ability to ask the right questions at the right time. However, there was a brief interaction in the car that just ruined the moment. Carol, Rachel and Jess all ended our meetings and discussions being polite and complimentary leaving you in a nice warm and fuzzy place, but Michelle was just about to change all that. We pulled up having completed the set up runs and Michelle turned to me and in a very enthusiastic tone and said "That was awesome man gimme five" Being very courteous I did eventually proceed with the high five but I think I may have inadvertently left her hanging for a moment as this really isn't a British thing and the abundance of engineering energy that filled the car a couple of minutes ago had gone.

It was a shame because at the height of our technical discussion on the adaptive shift strategy of the ZF automatic gearbox she was heading straight for number one on the laminated card of dreams! Having passed on all the important things about the car I dropped her off at the Winnebago to freshen up a bit and let her personal make up girl prepare her for the filming. I returned the car to Donnie for its own freshen up and make up and made my way back to the crew for some Hollywood type banter as you would.

When it came to the moment for filming everything went like clockwork. Ben took over the duties in the SVR with Michelle and I took to the wheel of the Ford Mustang with Al in the passenger seat controlling the video cameras.

The helicopter took to the skies to capture the aerial footage and Mary and Louise were busy organising the ground crews to get ready and close the desert road. We did a few high-speed runs to get some close up shots of the car and the helicopter did a number of close fly pasts to get a few different video angles. We then paused briefly to check all the footage and allow the road to open again making sure its users were happy.

I spoke to Ben who confirmed that Michelle was driving well and he was happy for her to continue up to 200mph. So, with everyone in a good place we positioned the car for the final run and advised the Department of Transport guys to close the road once again. The helicopter was back in the sky and they confirmed that the camera was rolling and it was time to go.

I stood in the middle of the road and watched as the SVR came to life and sprinted into the distance with the helicopter in pursuit. The noise of the exhaust was epic breaking the silence of the Nevada desert and I'm sure all the cattle spread for miles must have been thinking 'What the hell was that?"

Once the car was well out of sight we all got in the support vehicles and followed it back to the finish line where the helicopter was just landing and the transport guys were eager to get the road opened again as they had a handful of vehicles waiting to carry on north. Even with the whole spectacle of a gorgeous Jaguar SVR, a Hollywood movie star, film cameras and very low flying helicopter the onlookers in the queue were still disgruntled having been made to wait for fifteen minutes. One guy shouted out at us "Have you got nothing else better to do?" Well, actually mate no, I haven't got anything better to do because there is nothing better than this!

It really was a moment and I have to say even I was fully engaged with a lot of high fives and plenty of hugging.

We had achieved everything we wanted and done it safely. The whole team had enormous fun doing it and new friendships were born as a result.

Michelle was so chuffed with her run clocked at 201mph and was now a big Jaguar fan and at the end of the day she was more than happy to do a number of static photoshoots for all the team.

I gave her a big hug and thanked her for her support and paid compliments to her achievements over the day. She was blown away by the car and very complimentary in return to all the team which was nice!

Back at the ranch we said our goodbyes and she climbed on board her private helicopter back to Los Angeles and her fast and furious lifestyle.

For the rest of us it was cold beers in the ranch house with the family and then a return to Vegas for a night on the town to unwind, Jaguar style of course!

Sitting by the pool the next morning before our flights home later in the day I had time to reflect and feel very proud of what we had just completed, and it couldn't have been done without the absolute passion and dedication of the whole team. Mary and Louise had planned it all out in pain staking detail with perfect preparation and the support team of Andrew, Britt and Nikki were amazing. Then the lovely Juliet for all the UK PR support and bringing some Irish glamour to the proceedings. The production of the film by Tim Walker was excellent and captured by the legend that is Al Clark and all the epic photo stills were done by Anton Watts.

Special thanks go to our tame racing driver & instructor Ben for putting himself in the hot seat at 200mph and being a top bloke and to Donnie for looking after the car.

And there's more......Tempt media for the precision helicopter shooting, medical helicopter and transport helicopter. I can honestly say I have never seen so many helicopters for a Jaguar event.

There was also great support from MSS Ltd stunt safety coordinators, paramedics, rescue teams, refuelling team, fire support, water trucks, traffic control, Nevada Department of Transport and last but by no means least the Jaguar F-Type SVR and its Hollywood superstar Michelle Rodriguez.

I will remember you all as the 'A Team' and our own hashtag #whathappensinvegasisallablur

We were without a doubt Fast and Furious!!

# 19. Above the Clouds at Pikes Peak

I was reading an article one night about the achievements of Paula Radcliffe and in it she wrote, "Never set limits, go after your dreams, don't be afraid to push the boundaries and laugh a lot – it's good for you!" and that is just what I have done in my years at Jaguar, but on this one we pushed a bit further.

With the XKRS-GT there were no limits. I realised my dreams, I wasn't afraid to push the boundaries and we all laughed a lot that's for sure. It must have been good for us because here I was at Oulton Park watching motoring journalist and racing driver Chris Harris put our new baby through its paces for a magazine article and video.

As I watched from the pit garage, I was joined by some old Goodwood friends Don and Justin Law, who were watching the GT with a keen eye. We chatted for a while and got into a discussion about skunk projects and pushing the boundaries in engineering.

I told Don that I was keen to do something else and he mentioned Pikes Peak. We continued to discuss how cool it would be if Jaguar could do something special for the 'Race to the Clouds' and it left my head in a proper flat spin.

On the drive home I started to think about all the benefits of a one-off event and how they could improve future products and promote the division.

Some months later when SVO were under the leadership of Paul Newsome, there were a number of meetings set up to look at our future and what we could do to grow the brand, so I just had to throw my hat in the ring.

I proposed that we heavily modify an F-Type to take on the oldest race in America and compete at Pikes Peak where we could show off the abilities and technology within Special Vehicle Operations. Paul was as keen as mustard and told me to put a presentation together ASAP. At that moment I could hear the 'A' Team theme tune in my head or perhaps a flashback to the Blues Brothers when Jake and Ellwood decide to put the band back together.

My first port of call again was to Joe Buck for a quick rendering of a lightweight F-Type race car and he didn't let us down. I made a call to Javier for aerodynamic support, Pooky for chassis dynamics and my old mates from the Bonneville project, Dave Warner and Simon Stacey to help with the powertrain. With these guys in the bag it was over to Whitley for a skunky chat with our best vehicle build engineer Adi Smith, who very nearly exploded with excitement when I told him what we were going to do. The band of brothers are re-united again.

With the main players in place and a couple of meetings to discuss the specification for the car I got my best coloured pencils out and went PowerPoint crazy. I was yet again in my element doing all the research, putting a team together and hatching a plan.

The presentation was coming on well but I had a couple of things missing and one of those was a suitable driver, but I didn't have to look far. Over the last 10 years or so there was one name that came to mind for doing very fast hill climbs in a Jaguar and that was Justin Law. He virtually owned Goodwood with his epic driving skills at the wheel of our very own XJR9 that won at Le Mans in 1988. I called Don Law and reminded him of our pit garage chat and brought him up to speed with our plan which he was only too pleased to help with, and there it was again....The 'A' Team theme tune just popped up from nowhere.

Armed with a stellar presentation (if I say so myself), I stand up at the next review and present my latest skunk project detailing how we could showcase the attributes of our future cars in the latest cycle plan by competing and being successful at the 2015 Pikes Peak Race.
It was well received and after a few days of debate I was given the thumbs up to get a team together and make it happen under the guidance of our SVO Research team manager Dave Foster,

who was no stranger to skunk works having done the GT rear wing for us before moving up to his new position.

With the approval sorted I went up to see Don and Justin with the good news and they were as thrilled as I was. We had a very long chat about what we were going to do, and Don offered as much help and guidance as we wanted.

Justin was buzzing and started asking all sorts of relevant questions that I would have to get the answers to because if he was going to drive flat out to 14,000 ft. with sheer drop offs all the way up then he was putting his safety and life in the hands of the build team and I was comfortable with that all the way.

I returned to Gaydon for a download with the team and they all set about designing our hill climb challenger whilst Adi and I set about the massive challenge of losing weight, not that we were too porky or anything, but the car was.

In its base specification our donor car topped the scales near to 1850kg and there was just no way we could live with that, but we were lucky in that other projects had looked at weight loss and we managed to dig out previous weight walks to see what we could do.

On our extensive list we wanted to get as many carbon parts and panels on the car, completely strip out the interior leaving just the bare necessities (I bet you are singing the song now), remove and scrap anything that we didn't need (like the aircon), make a smaller fuel tank, swap glass for perspex and if we needed to, we could always get Justin on a low carb diet as well.

From Paula's quote at the beginning you should never set limits and with the weight of the car we didn't. Everything was considered during the strip down and re-build and we went after every ounce we could but then we had a large input of weight to consider that would make the task even harder with the addition of a roll cage. This wasn't your normal UK race cage, because after we read through the race regulations the specification for this cage was like something you would make for a military vehicle but I guess the nature of the race meant you had to have a substantial roll cage.

211

In our area we had a contractor called Chris Gradon who had many years in racing, supporting the World Rally cars and Touring Cars so it seemed perfect that we handed this baby for him to sort out leaving Adi and I to focus on the other million and one things.

With a copy of the rules and regulations we both sat down for a read and when we finished we looked at each other and thought the same thing.... 'Holy cow, we are going to need some help with this.'

A phone call or two later with 'The Don' and we have another team member called Alastair, who has worked with a few race teams and will pick up and look after all the race regulations for us. Adi is happy now that he can just focus on the build and it also allows me to run the rest of the project.

Things were now starting to move and gather pace and I was finding myself working all the hours I could just to keep up. I was busy most of the day on F-Type SVR development, fitting in the race car when I could during the day but as soon as I got home, I was straight onto the laptop working until the early hours of the morning. There is no overtime for this either, but I'm not concerned about that in the slightest. This is an opportunity that doesn't come along too often, and you just have to grab it and if you really want it to be successful then you just have to put in the graft. Luckily for me my team mates also have the same attitude and are all pulling in the same direction.

To be honest there is nothing better than the peace and quiet at home to do a lot of this stuff and the hours just get lost in some sort of time vortex, similar to when you get on the PlayStation with Grand Tourismo and the hour that you thought had elapsed had actually been about three hours.

With limited sleep I'm still up at 5:30am and off early to work to discuss the day's activities with Adi, and as always, we have various ideas about the project from the night before that we share and agree.

While we have been busy with the build, Don has completed and submitted the application for a start in the 2015 Pikes Peak

Race and has been contacted by the organisers. They are thrilled that we have entered and are excited to see a Jaguar on the start line. There are a few things we were unsure about with the regulations but the organisation bent over backwards to help us and we start to develop a good relationship with them which always helps.

During these discussions they ask us what race number we would like and I opt for 007 but with links to James Bond and Aston Martins we ditch that. I consider 00 for Dick Dastardly but it would be lost on the younger members of the team so in the end I decide to go for 152, which was the project code for the F-Type.

It is accepted and we are in but it only leaves us five months to get the car built so I need to recruit more skunk workers to the team which was going to cost me a fortune in coffee and biscuits (they are a cheap bunch though).

With lots of secret squirrel meetings lurking around the canteen coffee machine I first of all enlisted the help of Andy O'Toole who was the king of finding anything. He had a great background and a personal stash of useful booty in his man cave when it came to building race cars. He was then followed by Damon Fuller, who worked in Chassis who we needed to look after suspension, brakes, wheels and tyres. He used some of his previous race knowledge to guide us with our first ever carbon wheels, race brakes and slick racing tyres.

My next signing into the fold was Ian Parsons as our engine calibration man who would not come for coffee but when I mentioned Earl Grey I had him. Ian had done numerous calibrations for me on XFR Bonneville and XKRS and XKRS-GT and I knew I could trust him to get every ounce of performance for us.

For the gearbox there was only one man for the job in Roger Wardle, or Dodge as we knew him. He had been doing ZF calibrations for years and in order to carry over the original box we needed some funky magic doing to take the additional torque.

For race seats and restraints, I called on my XKRS-GT man Mark Twomey. No coffee required as he practically begged me to be on the project and as he also raced Global Lights in Ireland, he had good race knowledge and helped out on all sorts of stuff.

Over in Dynamics, Pooky had gone wild with the coffee tokens and enlisted Chris Spindler and Matt O'Hara to help him, Alex Jefferson to oversee the steering and Andy Gosling for everything on Stability Control Systems (SCS) and Traction Control Systems (TCS) with his sidekick James Rawlinson. It's a shame we hadn't invested in Costa coffee really because the shares rose rapidly over the next few months!!

With the UK team taking shape I had to start looking at what we needed to do when we shipped the car out to the USA and contacted Jaguar North America to speak to the lovely Teresa who guided us through all the shipping and temporary import regulations that would ensure safe passage from port of entry all the way to Denver Colorado.

That was a good start but now I had to get a team to help us in the USA and call in a few favours from my years in development and overseas testing. I contacted John Florida who ran the overseas test bases who agreed to help us take care of the logistics. He also agreed that I could talk directly with our operations in Phoenix with regard to resource but his caveat was that both he and his boss would both be in the USA for meetings in Phoenix during the race weekend and would love to join us in Denver. Done deal!!

I then called our base leader Craig Hopkins, who agreed to loan us the vehicle transporter and tools as well as two heads to drive the wagon and a support vehicle. He also accepted the challenge to find us some high-octane race fuel in the USA and get that shipped out to Denver (Top Man).

Our skunk project was now growing and we're working our socks off around the clock but we are just loving it!

We were well into the project before I realised we didn't have a project code that the team could refer to, and you just had to have one. Life in and around development lived by project codes and conversations were always started with "What project are you working on now?" and whilst 'Skunk Project' had a mysterious and somewhat secret military plane feel, we wanted our own identity so I decided on FPP15 which rolled nicely off the tongue. F for F-Type, PP for Pikes Peak and 15 for the year of the race. The team liked it, so it was logged onto the system and became a badge of honour.

I lost count on how many times I was asked what I was working on and in replying FPP15 there was always the response of "Oh, what's that then?" leaving it open for the obligatory response of any skunk worker to say, "I could tell you but then I would really have to shoot you."

Once we started using the project code around the business, word started to spread, and I was getting calls from all over the place wanting a piece of the action. The enthusiasm was fantastic and in the end I had to start handing out small tasks to keep people sweet.

With the clock ticking the pressure on the team was huge. Don and Justin were on the phone every day for updates and came down once a week for a progress update. We had a slot booked on the plane to fly the car over and we knew that failure was not an option.

Javier had bust a gut on the aerodynamics and came up with an amazing package. It had the biggest front splitter/spoiler I have ever seen, massive air dams over the wheel arches and a complete single piece carbon fibre underfloor and epic rear diffuser, creating our best friend 'Downforce' which we had loads of. When I saw the rear wing I was blown away. For those of us who remember drawing racing cars as kids and maybe as young adults you always had a massive rear wing on the car that made a statement of your intent. Well this was a statement. It kind of said "I'm a bad Mother F*****" and I mean business." It was humungous and we loved it. Javier told us that the data from the rest of the car dictated the size of the wing and that was good news!

The car was now also wearing a carbon roof, doors, rear decklid and bonnet saving a lot of weight.

Our perspex rear screen was in, as was the lightweight fuel tank and the whole car was taking shape.

The build was being completed at our Whitley workshops and the team had managed to get a secure area hidden away with key card access only, but that didn't stop the constant stream of visitors who had got word that something special was being born in the far corner of the shop. It would have been easy to just keep people out or turn them away but for the life of me I don't know why some people do that. My view was that we should let people take a look and get some enthusiasm and energy from what we were doing and on a few occasions the visits helped us out when engineers would come forward with ideas or knowledge they had from other projects.

There was most definitely a buzz in the place and how good is that?

A bit like my trip to Australia, I could write an entire book just on the build of this car but in order to get to the actual race I am just going to have to fast forward a bit. So, for those technical types that might be reading this one day, I apologise for missing out on all the automotive foreplay and technical erotica that would get your juices flowing but, for those who just want to get to the climax, let's do it.

After a particularly long day it is finally finished, and looks the business. It's raw, aggressive, and looks fast, which has to be a good thing, but as of yet, we haven't started it. With a mountain of anticipation on this special evening we are joined by our engine management guru, Dan Banks. The modern-day cars are controlled by loads of electronic control units and processors that individually are owned by each development area. Engine, gearbox, stability and traction control are the main ones but getting them all to talk to each other and then bypass the ones we have removed, as they were surplus to requirement, is all down to Dan the Man. He is flat out doing his day job at the track but that never stops him writing all the software and installing it out of hours for his mates and for his love of the brand and everything that is Jaguar.

After a lot of final checks, I get the honour of pressing the magic button.

Sitting there looking at the start button is the moment of truth, and my heart rate increases. I take a deep breath, keep my fingers crossed, and push the button.

The engagement of the button sends that all important cluster of electronic wizardry shooting down the wiring at 280,000,000 meters per second to energise a series of modules that co-ordinate the injection of highly volatile fuel into the five litre engine, and with the work of the devil himself, it ignites into a fireball of expanding gas, whose force transfers via the piston and conrod moving up and down thirty three times per second at idle to the exquisitely machined crankshaft, that with the flick of a paddle shift lever on the steering wheel, will let loose the 600hp through the drivetrain and propel the car like nothing we have done before.

Being in a confined area of the workshop I keep my hands off the paddle shift and opt for just a bit of right foot on the accelerator pedal instead.

Now I'm not sure if everyone does or feels the same thing but there are moments in your life when you just want to release all that emotion by just screaming or shouting out for joy? It could be the magic moment with your partner, the winning goal at a world cup, the birth of your first child or the moment of release after being constipated, but whatever it is, you must know how I feel when the car busts into life after being just an idea in the pit garage at Oulton Park, to a drawing on a sheet of paper followed by months of design, planning and build by the most dedicated team of Jaguar petrol heads in the company.

The noise is amazing. The straight through exhaust tells you that this is an out and out race car!

It is emotional, just like seeing your child being born, and there were tears for both!

The group hugs with the team were far manlier though and we were ready to Rock n Roll.

With most of the work completed we had one of our longest nights getting the geometry set up right with Pooky, Spindles and Matt, and with great pride we set the car onto the digital scales to record that our race diet had shed about 400 kg out of the car and our baby was ready to race. At 3:20am we drove her out of the garage for a parade lap around the Whitley site for no other reason than we could, and it made us feel good.

With a week to go before the flight I decided to go on a short break to Centre Parks in Norfolk with Danni, Olivia, Jess and our friends in the village, Darrel and Rachel. It was a well-earned break not just for me but also for Danni who had been an absolute rock during my months of dedication to this project and whose support gave me a lot of strength during some of the difficult times.

I left the final testing session with Justin, Adi and Javier to do at MIRA with some shakedown and track time to sign it all off.
On the first night of our break I got a call from Adi to say that the car had gone off the track and damaged the front end and undertrays and would need reworking asap, so like any good team manager would do, I cracked open some beers and delegated the task to Adi because Mrs Moore gave me the look that said if you go back now I will divorce you!!
By the time I did get back he had sorted it all out with a lot of help from Dave Foster and the Carbon Team!
After a brief re-test around the track and the thumbs up from all concerned the car was on its way and all that was left for me to do was pick a test team and then a race team.

With the car in the air I was back in Gaydon at my desk and on the table was a list of all the people who had supported the build of FPP15. If I could, I would have sent all of them to Pikes Peak just to watch the race, but my senior management team had told me that only a couple could go and support both the practice and the race.
The practice was crucial for us because as rookies none of us had ever been there and Justin would need to learn the 12.42 miles of road with 156 turns that went from 4720ft to 14110 ft.

The first name on the practice and race sheet had to be Justin, and Don had already advised me that he would just go on the race week, so those boxes were ticked. Don had also advised me that one of his race engineers was available for practice along with Alastair who had the most race experience. In the end we decided that the best person from the UK would be Javier as he knew the most about the aero package, rear wing settings, cooling and the vbox data logger. So that was it for the rookie practice week. For the race week Javier and Don Law's technician Rich would fly back to the UK so that Adi and I could join Don to fly out and be with Justin and Alastair.

Our team support from Phoenix would collect the car in Denver and stay there for the whole period. With the practice team now in Denver they set up everything for the additional practice week and all in all, things went very well. The times were promising as Justin learnt the course and the legendary Jeff Zwart, who was an eight times winner at the Peaks, stopped by to say that, not only did our car look good, he also thought that Justin was the best rookie he had seen in years.

This news was well received by the team in the UK but they were buzzing even more when we received the vehicle data and daily reports from Javier. We must have gone through gallons of coffee during this period because every chance we got we would meet up just to share our thoughts and our emotions.

After the last practice we had a report back from Justin saying that the car was amazing with lots of power and huge amounts of grip but any improvements in time would have to come from him as he needed to push even harder as his confidence grew. From the data, the only concern we had was the temperature of the gearbox near the final mile of the course, but there was not much we could do now as it was our time to go and take over the reins.

The day before I left was amazing. So many people came over to see us and wish us all the best and there was the general feeling that we just might have a chance of a top ten and maybe get up to about fifth if we could link all our best sections.

Paul Newsome took me to one side to wish us well and do the best we can, as we had already proved what we were capable of by getting the car built in five months with lots of new technology. He was as excited as we were and without his energy and support it would never have happened.

So, the three of us finally arrived in Denver and met up with Justin and the boys at our hotel where we were given the usual ten minutes to get unpacked and ready to go out for a meal in order to relax a bit after the flight.

With the elevator in use, Adi and I decided to sprint up the stairs to our level on the second floor. By the time I got to my room I was panting quite hard and was annoyed at myself for being so out of shape that I could only manage two flights of stairs. When I was ready, I went back down and we gathered in reception where Justin quizzed me on being out of breath. I explained my concern at my lack of fitness but a group within earshot started laughing and told me that Denver was at about 5600ft and it was the thin air causing the problems and not my fitness.

Adi then arrived with the same issues and we were amazed as to just how much it affects you. It's a good job we were not athletes though because the steak and beer we consumed that evening would have slowed us down a bit.

For our first day we had to take the car through scrutineering which we did without any issues and it was clear to see that all the organisers and most of the teams were pleased to have us there. Lots of people wanted pictures of the car and we were all enjoying a tiny portion of celebrity.

Having sorted all that out the team took the car away and Don and I went to race control to sort out all the passes and details for the next five days which also gave us the chance to meet the contacts that Don had made when he sent the entry. They were thrilled to see us as well and that felt so good as a rookie team.

Our schedule for three days was to complete the qualifying sessions that they broke up into three sections. The bottom of the course, the middle and the top and as luck would have it our first section was at the top without any time to acclimatize.

At around 5am the next morning the team started the climb to the top in the dark, which was a good thing really because had we seen what we did from our drive back down, we would have been far more nervous for our first run.

When we parked up on a very dark patch of land a few miles below the summit we had no idea of what lay ahead for us. We all had our hopes for the event, but we had no idea of what our competition was like or the conditions, and I also think we shared the same fears of what could happen if it all went wrong at this height.

We got out and started to unload the car and get our pitch set up and I walked quickly to the front of the truck to get some torches and jogged back which was a big mistake. Struggling to catch my breath I quickly realized that there really was a lack of oxygen at this height. We had all been given small oxygen cans to use if we felt unwell and without anybody looking I had a quick snort of the pure stuff and it was great. I walked back and continued without a word watching the others set about the normal race prep tasks at a very slow pace and then one by one they disappeared and returned looking as fresh as a daisy. It turned out that we were all doing the same thing using the oxygen, but none of us wanted to admit it or be caught doing it! My thoughts turned to the poor car that was going to try and suck in copious amounts of air shortly, and if we were struggling just how would it cope?

With everything ready I took a short and slow walk to the edge of the mountain road as the sun was just starting to rise. Looking out from my spot there was just a blanket of cloud that looked like the biggest duvet in the world and from just below you could see an orange glow trying to make its way through. The sky above the clouds was very clear and cold and the vast number of stars were being washed out by the light from the sunrise.

As the sun rose above the clouds you could see the road winding down from the top and it was just magnificent.
It could not have been more perfect. I was lost in my own space staring out to the east imagining what my family were doing right now as they watched the end of a day, with what I hoped was just a perfect sunset.

After all the planning and hard work, here we were on top of the world with our new baby ready to take on the 93rd running of Pikes Peak.

I heard some slow footsteps behind me and I was joined by Justin who put his arm on my shoulder and we both watched the rest of the sunrise from our amazing view point. No words were required but as we turned to go back to the car Justin said, "Shall we just rock this place or what?" At that point all I could hear in my head was "The Chain" by Fleetwood Mac and the rhythmic tones of the bass guitar plucking out the signature tune to the Formula 1 Grand Prix.

With the car ready to go Justin took up his place on the makeshift start line and the timekeeper waved his green flag. FPP15 burst into life and launched itself up the road with the noise of its very loud exhaust bouncing off every rock and boulder. It sounded epic through all the gear changes and crackled and popped during the downshifts and into the first hairpin. Although it was out of view, you could hear it change direction and come from the left to the right and upwards to the summit.

We were up and running and all we had to do now was wait for the timing sheets to be completed and posted before Justin and the rest of the group drove back to prepare for the next run. We had three runs to complete at the summit and then we would have to pack up and return to the bottom.

The timing sheets were looking good and the results of our early morning efforts were a top five place which we were really pleased about, although Justin still felt there was more in him yet.

The leader and veteran of our Time Attack 1 class was a chap called Jeff Zwart in an 1100hp Porsche and there was no way we could catch him with our automatic gearbox. It's a shame we didn't have time to do a manual version but that's how it goes. We were very happy with the morning's work so we packed up and headed back. On the journey down you could now see the drop off at the side of the road that we had missed in the early hours of the morning. Some sections must have had 2000ft drops and it was scary for sure. How on earth Justin would race up through here was beyond me and the lad certainly had some big Kahunas!

With the drive down we started to compare this place with the Isle of Man as far as road racing goes, but as this one does it on the side of one of the biggest mountains in the state it has to be the most dangerous by a long way.

To celebrate we found the most amazing restaurant run by a Mexican family who did the biggest and best breakfast burritos in Denver. Justin tucked in with the rest of the team and looked so calm, relaxed and happy and that was a good sign. If the driver is happy then the car is good and then the team will be happy. Bring on the next section.

The next day we headed back out to the mountain for the middle section at Glen Cove. The start point was further down which meant that it was far easier for us all to breathe. When we arrived we went to a local medical centre and managed to do a deal for a small oxygen cylinder that we then fitted into the car and ran a pipe into Justin's helmet. It worked very well on the top section the day before, but he felt that he was ok without it for the middle section.

It was a glorious day for driving with clear blue skies and enough heat for us to run a full set of slick tyres.

With an early slot Justin took to the mountain once more and roared off from the start point into a tree lined road. As he swung right out of sight you could still hear the epic noise of our exhaust as the car made its way up the road.

No sooner had the noise disappeared the next car arrived at the start and screeched off from the line. It was the American Porsche GT3 Race Team and they focused on the car as it twitched and hopped around the right hander.

When it had gone they walked back and stopped by our team for a chat. It turned out they had been watching our car with interest and wanted to know how come it went so flat round the bumpy first bend? I smiled and told them it was all down to a

223

special bag of magic pixie dust that my daughter had given me before I flew out. They saw the funny side and laughed with us. When they left, I turned to Don and Adi and said, "Did you see that? The Porsche Race Team were asking us why we were so good!" Hats off to the Dynamics team at Gaydon, you did a brilliant job!

With four runs completed Justin managed to cut his stage time down by fourteen seconds and was just holding onto third place in our group. Jeff Zwart was well out in front and in second by a whisker was a chap called David Rowe from Northampton in his 800hp EVO.

Things were looking good and Justin was still telling us he could find more time if he had more runs.

Whilst we were on this section we found some time to watch Rhys Millen in his all electric car, and although it was far too quiet for me it was unbelievably quick. His stage times converted into the overall win as well with a new record so we may have to return one day with an electric Jaguar maybe?

With things going well I was loving it. There was no stress, the team got on like a house on fire and we were doing everything together and I'm sure that translates to Justin when he is in the car.

During the practice runs Don took me to one side and said he would like it if only I spoke to Justin when he was in the car as I made him feel relaxed. If anybody wanted to say something or ask a question it went through me and it worked a treat. I have so much respect for Justin and his achievements over the years. He has won many Goodwood races and is one of the fastest men in the Festival of Speed. His times in the Jaguar XJR9 are insane and he beats all the top drivers. I went over to Le Mans and watched him win the Classic race in our XJR9. He was fast and very smooth with the car and I had a tear in my eye when he went over the line in first place.

Although he is totally focused in the car before the start of the qualification runs and he doesn't say too much, he does like some rock music when he is ready for it and Homero Mayol (our Phoenix technician) had put together some great tracks for the blaster and our favourite was 'The Devil n I' by Slipknot which soon became the team track!

Having finished day two of qualification what else is there to do than eat breakfast burritos again?

Our final day of qualification was at the bottom of the mountain which came and went without any dramas and we were holding another third place for the stage.

The car was just perfect, and the team were starting to get some interest from the others in our class and a few from the other groups. So many people came over to wish us all the best and they were so pleased to see a Jaguar at the event. It's just like being a father to a newborn baby when everybody tells you just how gorgeous it is and you pump your chest out feeling so proud.

Before we left the UK, I had contacted another Pikes Peak Legend in Paul Dallenbach who had also done some PR work for us with the New Range Rover. When we arrived, we made contact and during the qualification he would come down to see us and make sure everything was ok and offered us as much help as we wanted. It was a nice gesture that we took him up on when the road surface turned a bit damp and all we had was either slicks or wets.

Paul suggested that we cut some intermediate tyres and set us up for a visit to his garage facility in Denver. Homero and I drove over and with some guidance from his team we managed to cut a light tread into a set of slicks.

It wasn't just Paul either, as Jeff Zwart made us all feel welcome too and offered help and guidance from his many years of experience and wins at the event.

The whole paddock had a family feel to it. Everybody looked out for everyone else and I think it's because you are all racing against the mountain. Unfortunately, whilst we were on the bottom stage, the mountain claimed one of the family on the top stage in Carl Sorensen. He was a talented 39-year-old biker riding for Ducati on an 848 Superbike.

He was close to the summit and hit a bump sending him over the cliff and was killed in the fall.

It was a very sad day and a reminder to us all just how dangerous it was.

His family and team remained at the event and on race day, his paddock tent had a picture of him with some flowers and all who passed stopped briefly to pay their respects.

We had now reached our rest day, or should I say we had a day when we were not on the mountain but we were still very busy. Our team garage was in the local Jaguar Dealership in Denver who had bent over backwards to help us. They allowed to use all the tools and ramps and if we had any issues with parts, they would go off to find them for us. During our discussions with them it was mentioned that they had a really good technician called Mike, and it would be a great experience if he could be assigned to the car to help us, so we took him on board, made him an honorary member of the team and he joined us out on the mountain.

Being at the dealership was a bonus for us, and for them, as we had a lot of visits from customers wanting to see the car and take pictures with us. Whilst the team were working hard to prepare the car I ended up being the tour guide and storyteller of our escapades to anybody who wanted to listen.

With it being a rest day the whole town was having a car fest type of evening for Pikes Peak. The streets were closed off and there was a real party atmosphere. We had a blast with all the other competitors and all the locals who seemed to be rooting for us. As we walked up the high street that night, I watched all of my team enjoying themselves and it felt good. Adi was living the dream. He was instrumental in building us the most amazing car and his attention to detail was second to none. If he didn't like something or felt that it wasn't right then he just got on with it and made it right. We had worked on other cars before but this one had brought us very close and he was as proud as I was.

Don was as cool as a cucumber and was taking it all in. He was no stranger to this and had been around the world racing with Justin for many years and was a massive Jaguar supporter. Every time I looked at him there was a glint in his eye and a grin of inner confidence that told me something good was going to happen.

Homero, our Phoenix based technician was the main man. He always brought both a lot of laughter to the team and a huge amount of energy and I liked being around him. He was also the fittest bloke I have ever met with the biggest biceps in Arizona and that made me feel very safe in his company!

Alastair was a bit older than the rest of the group and was the wise old owl that kept us in place, but he was warming to the place and the event, but not our music!

Mike was your typical American down to earth biker who looked like a member of ZZ Top. You could see that he was enjoying being part of the team and when he put on one of our race shirts it was a badge of honour!

The main man though was Justin. He was taking it all in that night and enjoying every second. He had raced at Le Mans, Laguna Seca, Silverstone and Goodwood to name but a few but I got the impression that this was just a little bit special. When we bumped into Jeff Zwart we stopped and had a chat about life and Jeff patted him on the back and said that he was one of the best newcomers to the event in years and was looking forward to racing with him on Sunday. I shook Jeff's hand and thanked him for his kind words and wished him all the best.

We finished off the night meeting up with John Florida and Peter Richings who had flown over from Phoenix for the race. They had both supported us with all the logistics in the USA and it was great to have them on board with us as official supporters who then joined us for our biggest team meal and a few rounds of Samuel Adams beer. Happy days of the highest order!

Even with more than a few beers and the biggest steak inside me I found it difficult to sleep that night. My stomach was churning, and I could not switch off at all. I was starting to worry about the car even though it hadn't missed a beat during practice or qualifying and following the tragic loss of Carl Sorensen, I was also worrying about Justin should anything go wrong. I had spent an obscene amount of hours putting this project together and only now was I starting to wonder if we had done anything wrong and had I made any mistakes or bad judgements during the build.

I was starting to feel nervous and a bit stressed but then something absolutely wonderful happened. My phone rang and it was Danielle who had just got up at about 8am UK time on the Sunday morning of the race. She was face timing me so that I could also see my daughter Olivia who was just about to turn four. They told me that they had managed to find a link to a site that were going to do a live stream from the race and they couldn't wait. They also told me that they were so proud of me and the team, and that our whole department in SVO were right behind us.
Olivia ended the conversation telling me that she loved me and I was the best Daddy ever.
It was just what I needed, and all my worries turned into excitement.

Sunday the 28th of June 2015
It was an early start and pitch-black outside as the team assembled at the front of the main reception.
Everyone knew what they had to do and which vehicle they were in, so we all took our places and set off to the mountain. It was very quiet and I guess there was a hint of apprehension in the air but Homero sorted it all out when he said, "Gentleman, it's a special day so let's rock this place," and switched on the CD player with our favourite track from Slipknot cranked up to max volume. By the time we reached the paddock we were pumped and ready to go!

We arrived and set up our base next to Paul Dallenbach as he was part of our extended family even though he didn't care much for our choice of music either.
With the blaster turned down to number five (as it was still dark) we went about our business like a well-oiled machine.
Adi and Mike got the car out and set up on a ground plate while Homero set up the support truck and sorted through the wheels and tyres. Justin found a quite space in the Land Rover and put

some earphones on to relax and Alastair started to record cold tyre pressures and re-check all the fluid levels.

Don and I took a stroll to race control to check in and then attend a team brief for the day.

It was a cold morning but as the sun started to rise over the mountain ranges you could feel the warmth penetrating the trees of our enclosure and the increased light brought the paddock to life.

BBQ's were being lit already and the smell of burgers, bacon and sausage filled the air along with the overpowering smell of high-octane fuel and oil.

So this was it. The race was really happening, and we were here doing our bit for Queen and Country.

As I was walking back from race control I must have hit a good spot for reception as suddenly I had a mass of texts and WhatsApp messages arrive on my phone. They were all good luck messages from friends, family and work colleagues wishing us all the best for today and I enjoyed reading them all.

During the walk, Don and I stopped for a moment just to take in everything that was around us. The mountain stood tall in the skyline looking down on us as if it had just thrown down the gauntlet, the light breeze was blowing through the pine trees leaving a fresh smell in the air and the sounds of the paddock reminded you that we were here to race!

Just before we got back to our spot Don turned to me and put his hand forward to shake mine. No words were required, just a firm handshake, a look in the eye and a slight grin said it all.

With Justin in the car the team all moved towards the start line and I stayed crouched next to the open door.

I leant in and pulled down the restraints and buckled him in nice and tight.

As he adjusted his helmet I checked the oxygen system, and then checked his Hans system for head and neck support before passing him his racing gloves.

229

As we sat in line waiting for our start slot Justin looked over to the passenger door where we had put a map of the course on the door panel. You could see his head moving from corner to corner as he tried to visualise the course for the final time and when he reached the top, he just nodded that he was ready.

Up ahead you could hear the cars starting and then racing away into the distance.
With just four more in front of us it was time to start our engine. We had waited until the last minute to ensure that it wasn't too hot at the start line. Justin pressed the start button and FPP15 crackled into life once more.

As the cars in front started their run I continued to push the car forward until we met the rest of the team. I did some final checks and then looked straight into Justin's eyes. I gave him a nod and said, "Are you ready to rock mate?" His eyebrows lifted and I got a double nod. Homero brought the blaster to the side of the car and gave it full beans. I'm not sure any of the regulars had seen this approach before but they sure liked it!

This was it, the car in front left the line and the starter called us forward. The team pushed the car to the line, and I gave Justin's shoulder a tight squeeze followed by the biggest smile I had. As he looked at me, I considered all the motivational things I could possibly say but opted for "Do you think we should do Tilted Kilt tonight for some champagne?"
I closed the door and I'm sure I could hear him sniggering inside his helmet.

As the flag dropped, the fire breathing cat leapt off the line and howled around the bend and out of view and there was nothing more we could do.
We walked back down the paddock and gathered around the official timing screens so that we could see all the check point times. The tension was unbelievable, and the times were coming in for the first cars to reach the summit.
It seemed like ages before Justin's first check point appeared and as the time flashed on the screen, he was holding second place in Time Attack 1. Jeff Zwart was setting the pace as

expected but the jag was holding off the efforts of Dave Rowe in his Mitsubishi EVO.

We were pacing around the screen like expectant fathers when the second checkpoint arrived to confirm that Justin was still holding second.

The tension was unbearable, and we all had our fingers crossed as FPP15 was now entering the most dangerous sections of the course. My heart was pounding, and I had to take some deep breaths to calm down a bit.

At the next check point the two Brits were neck and neck but we could see that Justin was a bit slower than our qualifying times and we just hoped that everything was ok.

I wished we had a live video link to watch this because the timing screen was driving us crazy. The updates seemed to take ages and there was so much data to take in as well.

Looking at my watch I knew that Justin must be at the top, but the screen was just hanging on to the data. Jeff Zwart had finished and then the times appeared for Dave Rowe and Justin with the EVO just beating us.

Even though there were another nineteen in our class you could tell from the data that we were ok, and we started to embrace each other like we had won the race.

What a moment! As I grabbed hold of Don there were tears rolling down my cheeks. I hugged Adi and then Homero as if we had won the Formula 1 World Championship. It meant so much to us it was untrue.

Our euphoria was damped down a bit when the heavens opened and the racing was suspended. We made a run for it and set up base camp back in the transporter. As we sat there waiting for the rain to stop we had no idea that at the top Justin and a few of his group were in the thick of a snowstorm. When he did get back he told us that they actually had a snowball fight.

The rain soon cleared and brighter skies arrived for the racing to continue but we had to wait some time for all the cars to return from the top and back to the paddock.

The main road through the paddock was lined with a mass of people. Spectators and team crews were shoulder to shoulder and four or five people deep in places. As the cars and bikes arrived there was a mass of noise and each one received a round of applause. Some of the bikes did burn outs for the fans and they loved it.

Finally, FPP15 came into view and we all just lost the plot apart from Justin who as always looked as cool as a cucumber.

Back at our pitch she was switched off and Justin got out to share the moment with us. We had done it. We beat the mountain and made it onto the podium. It just couldn't get any better.

Eventually we headed over to the presentation tent and all sat as group in the middle. We cheered the winners from each category and gave warm applause to all the others and finally it came to our category. Time Attack 1 for production sports cars. As they read out the placings in reverse order we were just about to stand up when the master of ceremonies said, "In third place Justin Bell." I looked at Don and we had no idea what was going on. It had to be a mistake. I looked at a photo of the timing sheet which confirmed that Justin Bell was not even close.

I was fuming and after the presentation for Jeff Zwart I got up and stormed off to race control with Don in tow. I knocked on the door and went storming in with all guns blazing. Don was trying to calm it all down, but I had lost it. They eventually sat me down and a very nice lady went to retrieve the timing sheets. As they read through them the error must have hit them hard as they both looked a bit peaky and started apologising at a fast rate of knots.

The head of the event said that he needed to fix it straight away and we headed off to the paddock to get our trophy back but unfortunately the Justin Bell team had packed up at the speed of light and made off with our prize.

The organisers were now as upset as we were but assured us it would all get sorted out and the trophy would get back to its rightful owners as quickly as possible. After about 20 minutes they arrived at our spot with the news that Justin Bell had left the building with his Lexus crew and was heading back into Denver.

With the organisers being so apologetic my focus turned to Mr Bell. To this day I will not understand why he didn't just hand it back and say that it must be an error as he had the same timing data that we had.

With the race officials being so apologetic we decided it was time to pack up and head off for our own celebration instead with a fantastic team meal and a lot of beers!

The next day we headed up to the Jaguar dealership to tidy up and say a massive thank you to all the guys.

We boxed up all the spares and made sure that Homero was all loaded up with FPP15 for his return to Phoenix where he would arrange for its shipment back to the UK.

We did our final tour around the workshop and it was clear that we had become a group of minor celebrities. Lots of the guys were taking pictures with us and some had brought family in to meet us and see the car. It was overwhelming really as we had not expected any of this and in the end, we even started doing autographs and signing some of our used tyres that the workshop wanted to keep and put on show.

After the dealership we stopped off at race control to pass our thanks to their team for making us so welcome.

We told them not to worry about the missing trophy, but they were adamant that they were going to sort it out and get it back to us.

After race control there was one place left to visit which was the Mexican Restaurant for some morning after burritos.

The owners and their family were thrilled with our third place and we even had to get the car out of the truck for some team photos. Just brilliant!

We left Denver and flew back to the UK with our hearts full of pride. The whole team had done something really special and added a new line in the history books of Jaguar Cars but unfortunately, due to the release and launch of the Jaguar

Formula E, the PR Team decided not to run with our story which was a shame, although the impact within SVO was huge and it lifted morale in a big way and the lessons learnt doing the car were passed on for future projects.

It was some weeks later that Don got a call from Justin Bell to say he was sorry about the mistake and he would arrange to get the trophy to us. It never appeared but the organisers did get the record books changed and sent us a new trophy.

To all those who worked endless hours on this project and shared my passion for doing something a bit special I thank you!

**Adi Smith** – You managed the build of a great race car and became a brilliant wing man. Let's do it again one day!

**Javier Castane** – Your downforce made the difference my friend

**Pooky, Spindles and Matt** – Your vehicle dynamics set-up was a bit special…. ask the USA Porsche team!

**Mr Warner and Mr Stacey** – Power was and always will be everything. A great engine!

**Dan Banks** – For what you did with the electronics and software you should be in charge at Hogwarts. Just magic

**Damon Fuller** – You got us the right brakes and the lightest carbon wheels were sweet.

**Andy Gosling and James Rawlinson** – Another brilliant SCS package gents and NO you can't have your Electronic Control Unit back

**Chris Gradon** – The roll cage was awesome, and we were so pleased we didn't have to use it.

**Paul Newsome** – Your support and project approval was appreciated. Wish you were still working with us.

**Dave Foster** – Without your research budget and support none of this would have been possible.

**Andy O'Toole** – Master of part acquisition. You can't have anything back either. Possession is nine tenths of the law!

**Don Law** – What can I say dude. You sowed the seed and watched it grow. Your guidance and support made the difference. We will go back one day as we just have to win

**Justin Law** – Without you my friend FPP15 is just a race car. A very aggressive looking one to be fair but without you it would have achieved nothing. We provided you the tool, but it was you that made something with it, and like a chisel you carved our names into Jaguar History.

Finally, to my wife – you supported me from the very start and gave me the inspiration to follow my dreams. Without you behind me I could not have put in the hours to make this happen.
You're the best x

# Part 4 – Classic Jaguar Events

There are a lot of things that Jaguar do well and in the next five chapters I have covered the ones that I have been fortunate to have been invited on.

The Goodwood Festival of Speed is held every year and an event that is first on the calendar.
It allows us to showcase our finest cars and promote the brand in a very classic way.

The Mille Miglia is another yearly event for classic cars that gives us the opportunity to run our XK120's, Mark 7, C-Types and D-Types. It also offers us the chance to get celebrities and media into the cars and expand the brand image through their own network of social media.

The XF Launch in both Phoenix and Monaco is typical of a vehicle launch event where we set up base and bring in the worlds media to assess our new vehicles. Having been on the XF project for three years these two events were very special.
As with any of these launch events the car is always the star but the effort and attention to detail that our PR, Events and Communications teams put in is second to none.

The 50th Anniversary E-Type run to Geneva was so quintessentially British.
This iconic car has stood the test of time and what better way to celebrate than be on a road trip to the place that it was launched.

Reims de L'Elegance is the French equivalent of Goodwood but on a smaller scale.

This historic town and its famous Grand Prix circuit is the perfect setting to show off our racing credentials. It also allows us to return with our circuit winning D-Type and my XFR Bonneville Project, which just happens to be the fastest production Jaguar ever!

# 20. Glorious Goodwood

What can you say about the Festival of Speed? It's probably the best car show in the world when thousands of people spend three days in the summer sun watching Vintage Cars, Race Cars and Supercars taking on the challenge of the now famous Goodwood Hill.

Lord March opened his gates to the automotive masses in 1993 and he could not have dreamed how the event would grow into the massive spectacle it is today.

My first encounter with the festival was in 2006 when our PR team asked if I would support the event and provide engineering support for a couple of cars in the supercar paddock. It was a great gig for me to be honest! Being looked after for three days, looking after expensive and iconic Jaguar's and mixing with the VIP's of the motor racing world was a dream come true. I loved being with our cars in the paddock and talking to all the car enthusiasts.
There was such a buzz about preparing a car and then watching it fly up the hill and I always felt so proud when it returned as the masses gathered to get photographs.

During the course of the weekend I managed to go up the hill as a passenger with our legendary test driver, and dynamics guru, Mike Cross, it was mind blowing. When you watch the event on TV you have no idea how narrow the course is and how close all the spectators are too you as you approach the first corner. Then there is the elevation through and past the main house leading up to Molecomb, which is an off camber left hander that bites you if you get it wrong, followed by the flint wall and then a narrowing valley of hay bales to the finish line.

Mike had been doing this event for some years and made it look so easy as we drifted through and out of the first bend. He powered up the hill telling me when to brake and get the car sorted before the left hander at Molecomb and added a few words of caution should I ever get the chance to drive the course.

We finished our run and there I was at the top of the hill amongst a mass of gorgeous sports cars and some very famous drivers just hanging around for twenty minutes until we could drive back down to the paddock, little did I know but I would be returning again soon, but not as a passenger.

On the Sunday I prepared the cars for the two runs and was in my element. Sun was shining, cars looked amazing and everyone was having a great time. Our PR Manager at the time Ken popped over to have a chat and said that one of the journalists had dropped out of the Sunday drive so would I like to take the car up the hill……errrrr….let me just think about that for a nano second………..needs must I guess!!

I was gobsmacked, excited and nervous all at the same time but if I was going to be honest, I was busting a gut to have a go.

So, what's it like?  Well let me take you through it!!

I grab my helmet and settle down in the car whilst it's still parked up in the paddock. I know it's not a race, it's a demonstration run of the finest supercars on the planet but that doesn't matter one bit as the adrenalin still flows the same way. A whistle blows really loud and before you know it all the crowds are moved out of the way and you are beckoned out of the pits. Creeping forward to the next holding area is slow progress as the crowds are massive. Everyone is trying to take photos of the cars and I am feeling very proud and privileged to have my hands on the steering wheel of our new XK8. Although I'm forty two the eighteen year old in me feels the need to make some noise so I push the throttle hard to make some noise and the crowds love it.

I follow Mike in his XK, and we are escorted into another holding area where you can get out and stretch your legs. I get out and have a few words with Mike who gives me a few last tips and words of wisdom which still ring true today.

"Have fun, be careful and don't prang it or you will never be asked back again."

I decide to mingle and do a bit of deep breathing as the word prang in Mike's last comment made me nervous. Crikey…. what if you do prang it……..nothing like pressure then.

The holding area is an Aladdin's cave with Ferrari, Porsche, Aston Martin, Maserati and Lotus to name but a few.

Then there are the faces…. Jenson Button, Derek Bell, Chris Evans, Eddie Irvine, Mika Hakkinen, Jackie Stewart and Nigel Mansell and in the middle of all this is Dave Moore feeling lost and now even more nervous. It's just unreal!

There is then a call for all the drivers to make their way down to the start line and we all scurry back to our chariots.

As you line up for the drive down the crowds are really bearing down on you to get pictures of the cars and drivers and I start to feel relaxed and soak up some of the attention. I could get used to this!! The marshal beckons us forward and I watch the first cars scream away down the hill. Just to the left of the exit is the Jaguar Hospitality area so it seems only fitting to give it some, screech the tyres a bit and contribute to the automotive occasion or basically just show off.

I follow the car in front, and we turn around at the gates and then line up for the start which gives you another chance to get out and mingle for five mins. I walk up the line of cars giving a knowing look to fellow drivers. No words are required at this point, just a nod of approval at their weapon of choice and a smile that says, "Oh My God I'm really doing this".

If you have never been to Goodwood this really is a moment. The finest Supercars from around the world all lined up and ready to go. I'm not sure how many millions of pounds they total but I do feel blessed to have the finest Jaguar at my disposal. Engines start and the noise is epic as one by one they set off up the course.

As I get closer to the start line my emotions are fighting each other. In one corner I'm thrilled, excited and eager and in the other corner I'm tense, nervous and a tad apprehensive but before the referee can step in, I find myself at the front of the line. No time to think, traction off, engage gear, left foot on the brake and then full throttle. I know I'm not going to set the world on fire with this run and I'm 100% confident that I'm not going to get a race contract out of it but if there is one thing I can do it's a crowd pleasing burnout and what the heck, I'm here to please.

After about four or five seconds it becomes very real again as you emerge from the tree line into the first corner and you get to see the crowds. In the build up to this moment I have dreamt about going into the corner like some of the great rally car drivers with a Norwegian flick and an epic drift but in a nano second of seeing the crowds my own self-preservation applies a bit of trail braking and I opt for a more measured line around the bend. Now this may all sound a bit wimpy but let's not forget a few very important details here. It's my first Goodwood and I don't want it to be my last, the Jaguar hospitality tent is just coming into view on my left so I need to stay on the track and at least 60% of the crowd are all taking pictures or videos and I really don't want to end up going viral on YouTube.

Having negotiated the right hander though its full beans up past Goodwood House and under the bridge as fast as the XK can go, which is quite a thrill. As you start to crest the hill the packed grandstands make the track feel very small and you realise there is no room for error. I had some great coaching from Mike Cross and did all my braking before the crest and had the car set up nicely for the off camber left hander at Molecomb. It's a famous corner that has taken many. There are those who have just run out of talent and even some greats who just came in a bit too hot on cold tyres. Just ask Sir Chris Hoy or Andy Green (World's Fastest Man in Thrust SSC) or even British Touring Car Driver John Cleland as they are all in the Molecomb club.

Upon exiting the corner with my job still intact and the car in one piece its full throttle up the hill to the stone wall.

It's at this point where I feel I have some confidence being very close to the wall and I think it's a result of my training at MIRA on the No 1 circuit doing high speeds on a banked curve next to an armco barrier. I sweep through the right and then left with the stone wall coming very close to the door mirror and it's exhilarating. It's quite dark in this section but you slingshot out into the sunlight and finish the run between the hay bales flat out over the finish line.

As you slow down its time to breathe again and relax the grip on the steering wheel. It's all over in just over a minute or so but you have done it. There are no medals, just the honour and pleasure of driving a car as fast as you can for people to watch and enjoy.

The run ends when you park up at the top of the hill to wait for the rest of the field to finish and yet another chance to rub shoulders with your fellow drivers. This bit is now far better having done the run as you can contribute to a conversation like a true pro discussing lines and experiences from your drive and developing a bond with a limited group of individuals who have had the privilege to have driven at the Festival Of Speed.

As a result of a clean drive and being very enthusiastic I was given the opportunity to drive for a few years and it became the highlight of the year for both me and then my family for which we were all so proud.

I managed to drive the XKR, XKR75, XJR's, XFR, XF Bonneville, XFRS and XKRS-GT and I like to think that with each run I managed to get better and quicker. I think the XKRS-GT was my finest moment for a few reasons. It was a car that I took from concept to launch, it was a limited edition track car that the public were keen to see for the first time, and I also got to meet Sir Chris Hoy again after our Mille Miglia event. He had been invited to drive the car for one of the slots at the event this year before he then went on to join the Nissan Race Team.

Obviously the Festival Of Speed is just an epic event that I feel honoured to have been involved with and at the time of writing I hope to do again one day but I can't finish this chapter without mentioning a few other snippets of interest that make my normal life a little bit more exciting and funny.

First of all there is the drivers club. Well, this is an area of Goodwood where the drivers and VIP guests can sit to eat and drink and watch the event on Live TV. You enter through a

242

security gate and proceed up a red carpet to the main hall. When you arrive it's like being at the who's who of Motorsport that just blows you away. I absolutely love it and I'm not ashamed to say I get a massive kick from being there. It's so far away from reality you can't help but feel intoxicated by it. It was probably best in my first year as you become used to it over a few years, but you still sit there watching the famous faces walk by and sometimes sit down with you.

There was one afternoon when we were having an early tea watching the live feed when Eddie Irvine asked if he could join us. It was great sharing some time with him discussing some of the drives and talking about his time with Jaguar F1. Danielle my wife was blown away and could hardly say a word. After he left we both had the same view.... just a normal bloke who loves fast cars.

Meeting one of your Motorbike hero's in the toilet is another moment. Having completed a run up the hill I made my way to the driver's club bursting for the loo. At this point I must set the scene a little as this is no scabby nightclub toilet moment. Think along the lines of executive spa facility with wall to wall celebs getting changed and ready to drive.
As I'm standing there admiring the marble walls and fancy décor none other than Mick Doohan, the five times World Motor Racing Champion, steps up alongside. It's a weird, bizarre and somewhat surreal situation to be honest. I give him the courtesy look, which is then followed by the 'WTF' its Mick Doohan look, followed by the 'FFS' whatever you do keep your eyes forward and finish your pee look.
In an attempt to be uber casual I say "Gday mate how's it going" Why on earth did I say gday for Christ's sake? I mean, he is Australian but I'm as English as it gets. To make it worse I think I even added a twang of Ozzie to the accent. Thank god I remembered where I was as I'm sure my next trick would have been to shake his hand and tell him he was my hero. Whilst that episode could have turned so uncool, meeting another legend in John McGuiness (23 times winner at Isle of Man TT Races) was very cool.

In 2011 Danni was heavily pregnant with Olivia and was unsure if she could attend Goodwood but in the end she felt ok and didn't want to miss it. We were out strolling through the motorbike paddock when I spotted John. Luckily the bike riders have an easier time of it at Goodwood as most of the fans are crammed into the Formula 1 paddock and there does not appear to be too many die hard bike racing fans around. We wander over and I introduce myself and the wife and explain that Danni is due to give birth within the next couple of days and ask if he could sign our baby book. John is a great family man and was only too willing and said he loved his wife being pregnant. He asked if he could hold the bump and we had a great picture with him as a keep sake to show Olivia when she grows up.

The following year we managed to bump into him again and get a great picture with Olivia and when Danni was pregnant with Jessica, he suggested another bump picture to complete the collection with him.
A true legend and a great man. Thank You!!

Also, at the same 2011 event was Jay Leno, as a guest of Jaguar who would be driving up the hill in an XJR with me as a result of our LA drive in the XFR. It's always good meeting somebody again because at least you have a connection and something to discuss. During our stint on the track I mentioned that Danni was with me and very pregnant and he said he would love to meet her. I whisked her round to the paddock and he was a real gent who also signed the unborn baby book for us. I do hope Olivia loves this book in her later years!!
It was a lovely moment for Danni and some excellent kudos on Facebook!!

The list of Sporting Greats that I had the pleasure of meeting and talking to as a result of the Festival of Speed is long and distinguished, but I have to say that at the top of that list is the person who was the inspiration for the name of our second born girl. Dame Jessica Ennis-Hill was one of our guests at Goodwood in 2011 and I had to look after her as she was going to do a passenger drive in our XKRS. I had been a massive fan throughout her career and joined the nation celebrating her World Championship Gold Medals and other European Titles. For those who watch the TV sitcom 'Friends' there is an episode about the laminated card of dreams and all I can say is that Jessica was on my list and Mrs Moore accepted this as I know that Brad Pitt is on her list. Anyway, having been introduced I muddled my way through conversation like a ten year old sucking lemon sherbets. I had never been so nervous and after a couple of minutes I had to say to myself, David just stop talking. Having stopped, my wife still advised me that my mouth was open and should be shut. She was delightful and completely engaging and after our meeting there was no doubt in my mind that our second born would be called Jessica!! Her middle name is also Rossi but that's another story.

# 21. The Mille Miglia

The Mille Miglia (or a thousand miles) is one of the most famous road races in the world.

It was first started when the city of Brescia lost its rights of the Italian Grand Prix to Monza and feeling a little bit peeved Franco Mazzotti and Aymo Maggi, along with some wealthy families and friends, decided to stage their own race from Brecia to Rome and back again. The first of these races took place in 1927 and was won by Giuseppe Morandi in twenty-one hours and five minutes.

Over the years many great names have won the race including Tazio Nuvolari, Juan Manuel Fangio and Sir Stirling Moss, who incidentally won the 1955 race in a record time of ten hours, seven minutes and forty eight seconds at an average speed of 98.5mph which in 1955 was absolutely barking mad. You have to just think about that for a moment.......nearly 1000 miles through towns and cities in Italy at an average of 98.5mph is just amazing and he also beat his team mate by some thirty two minutes who just happened to be the legend that was Juan Manuel Fangio!!

The race was briefly stopped in 1938 by Benito Mussolini following the death of some spectators but returned in 1940 and ran until 1957 when a crash involving a Ferrari, that killed a number of spectators, put a stop to the race.

In 1977 the race returned but is classed as a rally parade (in name only) for original cars dating from 1927 to 1957 that follows the original routes over three days.

The race has become extremely popular with motor racing enthusiasts and celebrities, with the likes of Jamiroquai, Mika Hakkinen, Rowan Atkinson, Brian Johnson (ACDC), Jay Leno, Sir Jackie Stewart, David Coulthard and Sir Chris Hoy in attendance, and in 2010 I was given my chance to be part of the race as support for the Jaguar Heritage Team. I was both thrilled and proud to be called upon and that continued through to 2014.

Having continually badgered Tony O'Keeffe about the Mille Miglia for many years I was finally rewarded with this support role and my task, along with Richard Mason from the heritage division, was to be the support team to our very own 1956 D-Type and Reims twelve-hour race winner (393RW) and I have to say I was more than a little excited about it. The same car had also recorded sixth place at Le Mans in 1957.

The D-Type had a fantastic race history winning at Le Mans in 1955 and again in 1957 when it took five of the top six places and the car I was looking after was driven by Duncan Hamilton who won Le Mans for Jaguar in 1953 in the C-Type. I had seen the car many times at various events and I was aware of its history but when you are told that you will be looking after it for the duration of the Mille Miglia you become even more aware of its pedigree, its historical status and the enormous insurance value in the millions of pounds region.....around £15 million actually.

Full of enthusiasm, I set off from Stanstead on a regular Ryanair flight to Brescia along with a plane load of competitors and support teams for the short haul flight into Northern Italy. The Ryanair flight is the very last one before the race and I'm glad to be on it. I take my seat on the aircraft and as I take a quick look around I realise that it's full of Mille Miglia drivers and teams and the chap just in front of me is Nick Mason of Pink Floyd fame.

When we arrive in Brescia we are met by Tony and some of the team who have been out a day or so early to set up the base and take delivery of the race cars. The guys collect us in our support vehicles for the event which just happened to be the all new Jaguar XJ. The car has only just been launched and the chance to show them off at such a major event is too good to miss for the PR team.

Having dropped off all our gear at the hotel we head straight out to the indoor arena where the cars are held for scrutineering. The whole place around the arena is buzzing and when you walk in it just takes your breath away. It's just like an Aladdin's cave of automotive treasure. There are eight to ten rows stretching out for forty or so cars long with just about every prestigious marque there is. Within seconds I'm drawn to the most gorgeous and equally most famous Bugatti in racing history with the Type 35. The model won over a thousand races in its time including the Grand Prix World Championship in 1926 and the Targa Florio for five consecutive years between 1925 and 1929.

The car in its classic racing blue livery is gleaming under the lighting and I am just blown away. I had seen many of them in magazines, but this was the first one I had seen in the flesh and yes, I was salivating.

Having controlled all my emotions I lifted my gaze to see not just one but four of them and I was now in full sensory overload. We walked up and down the rows and there were Alfa Romeo 6C's, Mercedes SSKs, Bentley Blowers, Maseratis and Ferraris all over the place. You would have needed a day or so to just look at them all.

We continued through the rows to our own cars and as I walked along, I started to think about the total value of cars in one place and by my estimate it had to be in excess of £200 million pounds.

In 2010 we have six 'Works' supported Jaguars in the field, including three very noteworthy cars from the collection being NUB120, an XK120 from 1950 that won many races, NDU289 which was one of the fifty production C-Types from 1953, and then 393RW which was our 1956 race winning D-Type, and along with all the others, they all stand out like automotive royalty. It's so impressive and I could enthuse and write about it forever!

The one thing that will always be difficult to write about is the smell of classic race cars. The oil, tyres, leather, aluminium and years of usage is quite unique. They should forget the bottles of Chopard aftershave and bottle up the classic car smells instead. It's not quite as good as Castrol R oil that I used when I was racing motorbikes but it's quite close! Oooo and avgas (jet fuel) that's a bit special too if you know what I mean!

Although I'm like a kid in a candy shop, there's work to do and we knuckle down to getting the D-Type ready for inspection. Luckily Rich has done this before and we breeze our way through with all the correct paperwork and get it ready for the short trip back to the hotel.

It can be quite a long process, as it is Italy after all, and a lot of the teams will be there until very late, but the Heritage team have it all sorted and all that's left to do is savour the delights of a family owned restaurant with heaps of mushroom tagliatella and copious amounts of champagne. It's not surprising that the walls of the restaurant are decked out with Jaguar memorabilia as Tony and the team have been here a few times before.

On the morning of the race it gets a bit chaotic. Last minute panics with route books, stop watches, what clothing to wear, will it rain later or will it be dry, are the cars fuelled, what are the co-ordinates for the first stop and we haven't even been into the town centre yet for the pre-race viewing!

All the cars are expected to make their way into the town and park up for a few hours like a summer car show.

It's about a ten-minute drive but with all the race traffic it can take an hour. To make things easier the support teams decide to take control of the roads and ease the race cars through as best we can and its great fun. The first support car heads out with three race cars followed by another support car and three more race cars. When we get to traffic islands the first car pulls out and stops the traffic whilst the rest of the team go through. The following support car sprints forward and repeats the process at each island and when it comes to the traffic lights we get really ballsy and send forward two cars to block the traffic when the lights are red. The result is a lot of horn blowing but the race followers love it and cheer as we blast on past!

Having made it into the town centre you park up the cars wherever you can and just abandon them for a couple of hours while you saunter around and mingle with the masses. It's a bit like Monaco on a Grand Prix weekend except there about three hundred and twenty more cars to look at.

All of the bars and restaurants are full and the smell of pizza, lasagne, olives, parmesan and focaccia fills the air in the main squares, and for those not driving there seems to be an endless supply of red wine!

I stop to admire many of the cars and chat with a lot of the competitors who are more than keen to tell you all about their fine chariots if you have the time to listen.

It really is a wonderful occasion and it oozes class. You could never do this event in Nuneaton that's for sure!! I'm sorry Nuneaton, you are my hometown and everything but I have to face hard facts. When our time is done in the town centre it's another mad dash back to the hotel where each team can pick up where they left off with more last-minute panics. Everything is checked and double checked because once you leave, the Mille Miglia racetrack only goes south for the next day and a half and there is no chance to go back for anything.

In our team we have Bernard, who is our Jaguar regional director and his co-pilot who is a German journalist. We are the first to be ready but we wait for all the other guys and then head off en masse to the start line on the other side of town which is fun, especially as my guys want to do a stop off at the monastery as it's the 'In Thing' to do before the race and get some sort of blessing which I'm sure we need. As it happens, the place we went to was the automobile museum for the Mille Miglia to actually pay homage to the races which just happens to be located in the ancient monastery of St. Euphemia, so we might just get that blessing for all the speed limits we are just about to break! When we arrive it appears that a number of teams have had the same idea and they are all setting up for photo shoots and a quick look in the museum and yet again, this being Italy, there appears to be an abundance of wine flowing for those who need it. I wonder if they even realise that we have a race ahead with some very fast cars to drive for a thousand miles.

My guys wander off for some last-minute networking and I'm left with the car to take some artistic pictures. While I'm setting up I notice that one of the onlookers has a familiar face. He walks over to say hello and asks me about the car, and it takes me a few seconds to realise it's the singer Jay Kay of Jamiroquai fame and car fanatic.

We shoot the breeze about everything to do with classic race cars and I ask him if he would like to have a go in the D-Type, unfortunately he is contracted to another brand so it would not be good. Whilst we are chatting about the event I keep telling myself not to ask for his autograph as I need to be reserved and also look cool, but it all gets the better of me when I remember that my wife really likes this dude and I just have to get it for her.

We must have been talking for ten minutes or so when a stunning looking girl started to make her way towards us. JK looked at me, grinned and said, "Sorry mate this is a perk of the job." The girl promptly showed him a tattoo on her neck with his name in it and he loved it. He wished me well for the race gave me another grin and escorted the lady into the museum. If only she had come over and said "Hey, are you Dave Moore from Jaguar" it would have been both magical, and hilarious, but tragically she opted for the better looking, multi-millionaire rock star.

So, with everyone happy and a blessing thrown in for good measure we fly off up the road in search of the start line.

All the competitors have to line up in start number order so there is a huge amount of pushing and shoving to get through those who have been a bit keen to get in the queue.

Rich and I find our cars' place in line and take a deep breath. Engines are off and it's time to relax a bit. Cars are released every minute, so it's going to take a few hours to get through the field, but once you are underway it gets totally mad.

Our German pilots decide it's yet again time to mingle and disappear into the masses to take part in more social networking leaving Rich and I to push the car to the start.

As we get closer to the front you can hear the noise of the cars pulling away and making their way through the ancient city. The exhaust noise bounces off every wall, and as most of the cars have straight through exhausts the sound is epic, making the hair on the back of your neck stand up.

Rich and I are both fully immersed in the occasion, soaking up the atmosphere and wishing that we were driving this iconic Jaguar. Whilst we have been pushing it the crowds have gotten bigger around us trying to take pictures and in some cases, just touching the car which is fine. Moving forward we notice a group pointing at the side of the car and laughing and it bothers me that something is not right, but then I remember that my German colleague who is driving the car has his full name on the side door........Bernard Kuhnt. It has a ring to it for sure and fortunately he isn't! But he is late!

With only a few cars until the start, our dynamic duo return and we strap them in good and tight and do a final quick check that they have everything before we wish them good luck and remind them that we will be right behind them every step of the way.

As they head up to the start ramp, Rich and I run back to our strategically placed XJ and wait for them to pass us. After about five minutes you can hear the distinctive sound of the D-Type coming down the road, threatening to shatter every pane of glass in the street. They pass us and we pull out just behind for the start of the craziest drive ever.

The crowds are amazing and they spill out all over the road waving you on like you are some sort of hero. There is no loyalty to any brand or any driver, they are just rooting for everybody. Even when we are a few miles outside of Brescia the crowds don't go away. Every traffic island has a mass of flag waving supporters on it and there does not appear to be any space at the roadside either. Over the whole journey to Rome and back it's the same on every part of the route and we love it.

At a couple of points the volume of traffic going through some of the villages slows you down to almost a standstill and the fans are begging you to rev the car or do a wheel spin, even though we are in a support vehicle. Now I'm a fan of customer first principles and you never know if these spectators might actually go out to buy a Jaguar one day so I feel duty bound to give it a colossal amount of throttle and leave some tyre marks and a bit of burning rubber. It goes down very well with the masses and we get a huge cheer.

With the road stretching out now, we start to hit some decent speeds and make progress, allowing us to work out where we are in our route book as we shall be needing it as it gets darker and the line of traffic thins out a bit.

About an hour into the drive I get a phone call from Tony asking me what page we are on in the book and I confirm our current location to him. He replies "Oh good, we have a friend of Jaguar that needs some help. Can you stop when you see him and offer some assistance?" There was no mention of who this friend of Jaguar was, but he did tell us that his right-hand rear wheel had come off.

After a couple of pages in the route book we found our wounded friend and sure enough the wheel had definitely come off his wagon, and it was a big wheel too.

The driver came up to us and in an enthusiastic American accent thanked us so much for stopping to help.

We sorted out the car quickly and managed to get the wheel back on, but it was missing the centre locking nut which was the reason it came off in the first place. We had a quick look around but found nothing. Rich reckoned we should drive back for a couple of miles to look for it. It was like looking for the proverbial needle in a haystack but that blessing from St Euphemia at the museum was going to pay dividends for us because as the headlights shone out over the grass verges we noticed something glinting just off the road. We stopped and there it was, a centre locking nut for a very old Alfa Romeo 6C. We burst out laughing at our luck and both looked to the sky with thanks.

We drove back and our new friend couldn't believe his luck. We hammered the locking nut back on and lock wired it into position, and he was now officially back in the race. He thanked us again and insisted that we had a limited edition copy of Avatar as a gift. We accepted it and put it in the glovebox of the XJ before calling Tony to give him the good news. He was delighted that we were of help and also thanked us, but I wanted to know who this fella was and Tony replied "Its Jim",

Jim who? "Jim Gianopolus"……. well who is he when he is at home? There was some muffled laughing on the phone before Tony told me that Jim was in fact the Chairman of 20th Century Fox which explained the copy of Avatar!!! If I had known this I could have tried to blag my way into a new film as an extra…drat!

We get back on route and head for Bologna and as the sun starts to set there is a familiar sound from the rear and in my mirror I can make out the gorgeous silhouette of another D-Type. I move over a bit and the Ecurrie Eccosse Blue racer blasts past and gives us a friendly wave.

The noise is epic, and we listen as it pulls away down the outside of the traffic. Bellisimo!

We are now flat out ourselves trying to catch up when the phone rings to say that the guys had gone off the road and my heart sinks. They tell us that all is ok but are keen for us to get there ASAP.

When we arrive we can see Bernard waving to us and we pull up. His co-pilot was driving and had lost it coming out of a bend and slid off the road. In doing so it managed to go through a twelve-foot gap between some concrete and a metal fence missing both and then slid to a stop in a marshy field without any damage at all. We recovered it back onto the road, adjusted the underfloor brake scoops and it was ready to get back in the action. Somebody on high was definitely looking down on us and that blessing we had in Brescia was really paying off now!

We finally made it into Bologna in the dead of the night to find our D-Type was already tucked away and our pilots were already in bed so it was left to the two of us to raid the drinks cabinet for a cold beer and then get our heads down for the early start to Rome.

It was a gorgeous morning and we were up really early to check over the car and make sure it had plenty of fuel. Being an ex Le Mans car, the tanks are huge but the mpg is very low, so we have to keep a check on it.

The run today will see us continue south through Imola, San Marino, Rieti and into Rome.

We head out from the car park and get straight to business with the D-Type at full throttle and the XJ in tow. It's just fantastic watching and hearing the Le Mans car when it's flat out. In 1957 it must have been amazing to see it on the Mulsanne Straight at over 170mph flying past everything in its path. Even now some fifty-three years later it's still doing a flipping good job at it and at times I'm struggling to keep up until we hit the traffic and the climb into San Marino. With old race cars you have to be patient and have a modicum of mechanical sympathy but unfortunately the boys pushed it a bit too hard and we could start to smell the clutch.

At the top of San Marino we all pulled in for a coffee and Rich had a check of the car and it wasn't good news. The clutch had quietly passed away on the climb and the poor old girl was going to have to make her way back to Brescia in the back of a truck. It was such a shame and my heart sank.

I was loving the drive and the event but without our main focus the edge had gone.

Bernard and his co-pilot jumped into one of the other support vehicles as they were desperate to finish the route and Richard and I decided to drive as the sweeper vehicle and back up any of the guys if they had any issues.

My allegiance though had now switched to the C-Type with Mike O'Driscoll and I would be cheering him on from the comfort of our XJ.

After San Marino the roads open out and we are just flying to keep up with the team. We overtake most of the cars in front with some gusto and at one point we out gun a police car and he applauds as we pass.

Having been driving all day I'm really getting used to the total disregard for every signpost, traffic light and non-event vehicle that might slow us up. I do however keep a reasonable gap from our race cars as nobody wants to see a brand new 2010 model XJ tucked up the boot of a competitor. People need to get photos as well so I always maintain a good distance, unlike the Mercedes teams that seem to think the support car must be touching the bumper of its race car. Muppets!

255

As the spare support car, we decide to press on in front to do a recce of the next checkpoint and get some coffees ready. We pulled into a gorgeous little village and parked the car. Right next to the stop was a proper Italian café and the smell of espresso was to die for. We ordered some doubles and sat down outside the shop next to a lovely old lady. When she realised we were English she made conversation with us and talked about how beautiful the classic cars were and that they were from an era that she remembered fondly. She asked how we were enjoying Italy and the race and I told her it was an absolute dream to be part of the occasion.

She finished her drink and then holding my arm she told me to take good care, enjoy the experience and wished us all the best. As she walked away, I watched her as she admired both a gull wing Mercedes and a Jaguar XK140 before she crossed the road and wandered out of view.

The waiter brought us the bill and asked if we had a good conversation with the lady. I said she was absolutely lovely, at which point he told us that the lady was an old actress called Gina Lollobrigida. I knew the name and I remember some of the old films from the 1950s and 1960s where she took star billing as a major sex symbol.
I smiled and was thrilled to have been in her presence for a short while. Just like the Mille Miglia she was a class act.

As we got back into the car I said to Rich, "This is just bonkers. We are racing through Italy with no speed limits having the most fun you can with your clothes on. I have met a rock star, the Chairman of 20th Century Fox, a Hollywood Film Star and recovered an iconic £15 million pound race car from a field in the last twenty-four hours and we are only half way."

After the caffeine break and check point cards stamped, we were off again for the final run into Rome.
It was getting dark now so I was expecting the run into Rome to be a breeze but how wrong could I have been. It was utter chaos. Solid traffic everywhere and some three hundred and

256

seventy or so race cars and their support vehicles trying to find Parc ferme and the overnight stop. The whole place was buzzing and then all of a sudden, the cavalry turned up in the shape of about thirty or so motorbike cops and we were given a police escort through the chaos.

The group we were in had about fifty cars in it, and we were making a whole lot of noise. The crowds on the streets were cheering and waving flags and hundreds of race fans following the event in their cars were blowing their horns as we rolled through. I was gobsmacked because it was about 11pm and the scenes were something I would have expected if Italy had just won the World Cup. How I didn't get any panels bent on the XJ must have been down to our new friend St Euphemia and the Mille Miglia blessing.

All the cars had made it through and considering that they are all from the 1950s it's amazing that they are still performing as they do. The support teams stay in the car park and carry out all the checks and prepare them for the early start and it's about 1:30am in the morning before we all make it back to the hotel. We are still buzzing with adrenaline so nobody is off to bed yet. I can't imagine what it must be like to be driving these cars as a competitor, as it must be amazing. Even just being the support team is enough to put us on a high!

I bump into Ken McConomy and Nigel Webb who are in Nigel's C-Type. Ken who is our PR Senior Manager is loving every minute of the event as is our Chairman Mike O'Driscoll, who is in our own C-Type. We manage to find a load of cold beers and finish off the day sharing our stories from the last five hundred miles.

Before I go to sleep, I lie back for a quiet moment to just take it all in. How on earth did I make it onto this event? I'm in the middle of Rome (I think) on one of the greatest road races in the world looking after some of Jaguar's prized assets whilst I'm driving its flagship vehicle the new Jaguar XJ. I guess it's a combination of knowing the right people and being in the right place at the right time or I was just damn lucky.

I'm absolutely shattered though and with the help of the beers I eventually fall to sleep fully clothed.

The alarm goes off at 5:30am and I have not got a clue where I am. It must have been 2am when I passed out and I really need more than three and half hours of sleep. I open the curtains and Rome is very much awake already.

I meet Richard and he looks as bad as I feel. We grab breakfast on the hoof and within twenty minutes the Jaguar Team are back in the race to Brescia which will take all day and some of the night.

I'm looking forward to driving some of the route back as we head through several stunning places. The early part of the route takes us through Radicofani then north to the gorgeous city of Siena, which is famous for the horse racing through the streets and became famous in a James Bond movie. After Siena, it's a short hop to Firenze and then the epic Passo Della Raticosa high in the Tuscan mountains. We then fade to the east back towards Bologna before going north to Fiorano, Parma and the final leg into Brescia.

The scenery during this section is just how you would imagine Italy. The roads are lined by cypress trees, the fields are a wash of olive groves and vineyards, and the old stone farmhouses and buildings, with their deep red slate roofs, are straight from the oldest oil paintings. It reminds me of some of the scenes from Gladiator. It's stunning!

The Mille Miglia cars fit in well and at times the scenes in front of us are just like you have been transported back in time. It's a shame that we don't have the time because it's a perfect setting to do some black and white photography.

Anyway, back in the race we are now heading into Siena and the support cars are separated from the race cars and we make our way into the main piazza where all the cars will pass through a check point and parade around the streets.

Siena is one of Italy's most famous cities where the narrow streets are flanked by its ancient medieval buildings and the centre that we are in is the Piazza del Campo with a Gothic town hall and the Torre del Mangia which is a tall 14th century tower. The piazza is also the home to the famous Palio horse race which takes place twice a year.

Richard and I have enough time to wander around the piazza and take it all in. There are hundreds of people in the centre awaiting the cars and it isn't long before the first batch are released through the streets. You can hear them coming a long time before they arrive. The noise bounces off every building as they make their way through narrow streets and then to my right the first cars appear and drive up to the checkpoint.

The crowds flock to line the route and they cheer on each car as if it was their own. Every car and driver is a hero for just being on the event. The smallest little Fiats get the same applause as their bigger and faster legends.

We watch a few of the cars going through and I end up talking to a very pleasant Canadian who tells me that he simply adores the old Jaguars and would love to get an XK120. We tell him that we are one of the support teams for Jaguar and he is blown away. The look on his face when I got him to pose by our C-Type at the checkpoint was priceless. We exchanged email details so that we could send each other photos and I also donated my Jaguar Heritage jacket to him as he was clearly a big fan. It wasn't until later after the race that I found out that he was one of the head honchos of AT&T in the USA and was on his way to Monaco to watch the F1 race as a major sponsor. Hopefully he managed to get an XK120 and remains a loyal fan to the brand.

We leave gorgeous Siena and the focus is now on the final few legs back to Brescia.

People are still lining the roads everywhere we go and in some of the villages you have to slow right down as they form a single file through. With our windows open we both put our hands out for dynamic high fives but after a hundred metres or so they become quite sore but it's a hoot.

As the cars head into Bologna we decide it's time to take a detour and leapfrog the field to get back into Brescia ahead of our race cars. This will give us a chance to get a good parking

259

spot and find a decent space near the finishing line to get some photos.

We follow a couple of old Ferrari's through Parma and the scenes were mind blowing. Massive Italian flags waved, and the crowds went wild. They are just so passionate and loyal to the home brand and I love it. I even find myself blasting the horn to join in the celebrations.

We leave Parma with huge smiles and wonder what it will be like in Brescia even at this late hour. We have completed nearly a thousand miles in two and a half days with about six hours sleep and we should be shattered but the adrenaline and the crowds keep us going!

The final couple of miles into Brescia are breath-taking. Just about all of Italy have come out to applaud every single car back. Even our support car with its event stickers gets cheered on.
In front of us, cars have fixed fireworks and flares to their chariots and ignited them for the final stretch and three of them go side by side leaving smoke trails of green, white and red.
As we approach the final island where the race cars go straight on to the finish line, we are sent to the right and find a suitable place to park up and I find myself getting quite emotional about the whole thing. I'm about to give in to my feminine side when Richard gives me a massive slap on the back and says, "Come on you old wuss, there is always next year."

We fight our way to the finishing line and await our own heroes and legends and cheer very loudly as each one is introduced to the crowd. Our cars finish mid table, but nobody seems to care because it's not about winning, it's about finishing with a smile on your face.
Tony O'Keeffe manages to find us and in his usual style hands us a bottle of champagne and some glasses and we head off to share the moment with our teams.

When they are all back, we head off to our friendly hotel to put the cars to bed for a well-earned sleep and the rest of us head for the restaurant which has become a bit of a tradition.

It may be 1am but the family have stayed up to look after us all. Mushroom tagliatella by the barrow load and an endless stock of champagne, cold beer and red wine.

We had an epic night exchanging stories and sharing our precious moments. We started the event as six individual works teams with twelve competitors, nine support vehicles, eighteen support engineers and five PR and logistical support guys but ended it with many more having invited several of the private Jaguar teams into our family circle.

What an event. I have included this chapter in the 'Classic Jaguar Events' but it could have gone into the 'Events of Biblical proportions.' There really is nothing like it on the planet and hats off to all those who really raced it from 1927 to 1957. We managed it all in about fifty-four hours but Sir Stirling Moss in 1955 set a record by doing it in an amazing 10hrs: 7mins: 48secs at an average speed of 98.5mph and you just have to stand up and applaud that.

He may have even set the record in 1952 had his Jaguar C-Type not had steering issues.

I was thrilled to be invited back to support the Mille Miglia for another five years and there were some amazing moments worth a mention.

Returning in 2011 we decided to showcase the XFR as our support vehicle and I had the pleasure of Stuart McEvoy from the Heritage division as my co-pilot. During the race we managed to find ourselves behind two brand new Ferraris and they were pushing them hard through the countryside. What they didn't expect was a Jaguar Saloon sitting on their tails for a few miles. Realising that they were not quite as brave or as capable as their cars, I decided it was time to let them see the new cat leap ahead of them. Before I did this though I got Stuart to grab his car magazine and pretend to be engrossed in reading whilst I did the overtaking. With a nice bit of road ahead the 550-horse powered V8 supercharger flew past in a blaze of glory as I shouted "Read the badge!"

Having shown them what this new Jag was all about I looked over at Stuart, who quite nervously popped his head out from the magazine and said, "Are we done yet, I need to be sick."

In 2012 I was asked to support a Jaguar VIP rather than our Heritage Team and my co-pilot for that year was Andy Lowis from Powertrain. We were assigned to a chap by the name of Paul Polman who was the CEO for a company called Unilever. I called my wife to find out who he was, and she burst out laughing and said, "You don't know who Unilever are?" "Nope," I replied. Turned out that they were one of the biggest companies in the world but as they didn't make cars or planes how was I supposed to know?

During the race Paul and his Swiss watch designer Co-Pilot managed to crash on a mountain pass and they thought the race was over, but Andy and I spent most of the night repairing the car and had it ready for the start the next day. They were delighted to be back in the race and we had an absolute blast, even though we both put on a few pounds on a diet of pringles and beer for three days!

In 2013 I had another challenge when the company decided to run a fleet of the new F-Type convertibles alongside the Jaguar Works Teams as well as a great list of celebrities.
We had actor, Daniel Day Lewis (Lincoln & Last of the Mohicans), Jim Gianopolus (20th Century Fox), models, David Gandy and Yasmin Le Bon, Le Mans winner Andy Wallace, journalist, Chris Harris and the Olympic Champion, Sir Chris Hoy.
It was another epic Mille Miglia and for me the highlight was meeting Sir Chris Hoy who was an absolute gentleman, apart from the moment he patted my belly and advised me to do some bike riding and lose a bit of weight. I also had a blast with the journalist Alex Goy who was in one of my convoy of F-Types.

I met up with Chris again at Goodwood on a couple of occasions as he kicked off his car racing career. He drove my XKRS-GT up the hill one year and then followed me up the hill the next year in his Nissan GTR but fell off at Molecomb.

I was teamed up in my car with Sarah Brautingham from PR which meant that I was going to be driving the full one thousand miles while she kept track of all the cars and people throughout the race. By the end of it she knew my wife's birthday, all my family's birthdays, what food I loved and who all my favourite Manchester United players were. Well played lady!

In my final year at the Mille Miglia we went out in style. I was put in charge of several F-Type Coupes that were going to race alongside the historic cars with journalists in each one and the Jaguar Works Teams were going to be driven by a whole host of celebrities. My co-pilot for the race this year was the lovely Sam Adams from the events team, who was also going to try and conduct proceedings whilst being my passenger.

Martin Brundle and Bruno Senna of F1 fame were going to drive our D-Type in hope of winning the event and then we had Jay Leno, Ian Callum, pop stars Eliot Gleave and Milow, Jodie Kidd, Special Forces hero David Blakeley, Jeremy Irons and AC/DC front man Brian Johnson.

Wow, what a line up!

It was just so cool I don't know where to start. Let's just go through the list as mentioned above.

Martin Brundle and Bruno Senna were a class act in our D-Type and I thought they would have a chance of winning it but unfortunately the timed stages got the better of them. In terms of raw pace, they were up there!

During the race Martin shot past my F-Type and I decided to tail him for a while. It didn't last long and I had to let him go.

Jay Leno was teamed up with our very own legend in Ian Callum and they were having some fun both on and off the road. I bumped into them at one of the lunchbreaks and it was hard to get them away from a splendid lunch and back into the race. They were both enthusing about the Italian cuisine and fine red wines as well as comparing notes on many classic race cars.

We had two pop stars, but I only really spoke to Eliot Gleave (stage name, Example). He had joined us with his wife Erin who was also a model and actress. He was definitely a petrol head and loved the race.

When we finished in Brescia we went out for the night and had our own party with a DJ. The highlight of the evening though was the DJ playing one of Example's biggest hits whilst he was dancing on the floor with us. The DJ was completely oblivious to this moment.

Jodie Kidd was driving with us alongside her then fiancé David Blakeley in a rather splendid XK120. As soon as I saw them on the team sheet, I knew we would be having an interesting chat. At the hotel after exchanging pleasantries I told them that I had something in common with both of them. My link to Jodie was that I had once ridden her Argentinian polo horse when I had a day learning to play polo. She had passed on the horse and remembered it well.

The common link to David was a lot closer to home though. The ex-Special Forces Captain, and author, was born in the same place as I was, The George Elliot Hospital, Nuneaton. When I pointed this out to him he was amazed. He had never bumped into anyone from Nuneaton since he left when he was young.

His two books 'Pathfinder' and 'Maverick One' are amazing and well worth a read!

When it comes to rock superstars and legends you can look no further than Brian Johnson from AC/DC. Brian was larger than life at the Mille Miglia. When he arrived, I met him in reception and took him through to the bar to meet up with the team. Within minutes we had beers and were talking everything about cars. To prove his point on being a car freak he promptly pulled up his boxer shorts to reveal that they were from the world famous Nurburgring racetrack.

He was immediately the centre of attention at the bar and was the ring master for the evening's entertainment.

During a discussion with David Blakeley, the ex-captain revealed that when they did high altitude low opening parachute drops behind enemy lines they used to jump out to the sounds of Thunderstruck by AC/DC. Brian was thrilled to hear this. It was a long night at the bar with Mr Johnson and we had a blast.

During the build up to the start of the race it was evident that every AC/DC fan in Brescia was out to see their hero and we decided that we would have to sneak Brian into the car at the last minute. Two of us were detailed to look after Brian and we took him the long way round to the start through the narrow side streets. As we approached the main start line we could see the car and the massive crowd around it. I hid Brian and his wife in a doorway until the car was a lot closer to us. When it was just opposite I got the nod from the team and we ran to the car and strapped him in before anybody knew what was happening. Once in the car all hell broke out and it got a bit scary. Fans, photographers and journalists were all over the place and we could hardly push the car along. Just BONKERS.

During the race it was the same at every checkpoint and it got to the point where we took Brian out of the race car a few miles before each stop and hid him in a support vehicle until the car was back out on the road.
He finished the event and we had a proper Rock n Roll night out whilst he was there to celebrate.

My personal favourite moment was taking the F-Types through a massive tunnel with the intention of making a lot of noise. With a vehicle flight pattern to match the Red Arrows we banked into the tunnel and opened up the F-Types exhaust reheat. The noise was breath-taking, and we maintained it for the entire length of the tunnel.
As we approached the end there was a different sound mingling with ours, and not born from a Jaguar.
We pulled up at the toll lanes and then realised that two rather gorgeous Pagani Zonda's had tagged on with us. They came over to praise the noise from the Jags and as excited Italians do, the expressions were amazing. "Semplicemente stupendo. Bellissima, Bellissima," Just gorgeous. Beautiful, Beautiful!

265

The decision to use the new Jags was the right one!

So that's the Mille Miglia. Over those years I could have written a book just on the races, as each one was a new adventure. I made a lot of new friends and new connections that would end up helping me out years later with this book.
We pushed ourselves to the limits to ensure that our cars made it home and our drivers had the best time ever.
It's the most insane motor race in the world and for that reason we love it to bits and for me it was an honour to be there supporting our brand.

Grazie Italia.

# 22. XF in Phoenix & Monaco

At the end of every development programme you go into the launch phase where you release the new car to the press for their first assessments. In the good old days the drives were conducted in the UK from MIRA but with XF it was going to be a bit special.

The PR and Events team had decided that the early reveal would be in Phoenix, Arizona, and the full press launch would be out of Monaco. As the Vehicle Office lead for the programme, I was made aware of the drives quite early and was extremely excited at the prospect of being asked to support both events.

Throughout all the testing I lived and breathed everything about the XF. I spent weeks and months going through every single detail of the development plan to ensure that our engineers got all the time they needed on the cars. I worked alongside the build engineers when they created the prototypes and made sure that every issue was reported and fixed, and towards the later stages I worked alongside PR to make sure all their requirements were met for the upcoming launch drives. The development phase takes nearly two years and during that time I was supported by some great engineers and friends. Matt Eyes and Michael Rodley were my wingmen in the Vehicle Office, and we all worked for Graham Wilkins and Andy Whyman under the programme leadership of the best Chief Programme Engineer, Mick Mohan.

Mick was a legend at Jaguar having worked his way through Body Engineering to get to the the ranks of Chief Programme Engineer.

I recall in one project meeting when I was asked to do a presentation on the status of the development testing, and having been questioned on some of the details I started to dig a hole and realised there was no escape route. Mick intervened and said he would like to review it after the meeting and that we should move on. I sat back down feeling a bit flustered with sweat running down my back. I was pleased that Mick had saved me from the vultures though.

I went to his office later in the day and was greeted with a smile and a friendly mates chat about the current testing. Having been given a cup of tea Mick then gave me the bollocking, quickly followed by some guidance on what I should have said and rounded it off with some praise that I was in fact doing a great job. For me that's the measure of a great leader. He made sure my bollocking was in private, he coached me and then made sure he lifted my spirits with some praise. After that episode I would have walked on hot coals for him!

As a result of his leadership, the XF programme had a family feel to it, where everybody was pulling in the same direction, for the same goal, and were having a lot of fun doing it along the way.

In August of 2007 I was away testing in Phoenix and I got call from the PR team to ask me if I could look at sorting out a drive route for the upcoming press drives. With the help of an old Phoenix work colleague, Pete Perrone, we spent a couple of days looking for the best roads and stop off locations. I sent a draft of the suggested route back to the UK and four months later I was back working with the team getting the cars ready for action.

Our primary location was a stunning place in the wealthy area of Paradise Valley. The cars and events team were based in a humungous tent on part of the even more humungous car park. I joined up with a number of old mates from Whitley including Andy Smith who worked with me in Vehicle Testing and all of his technicians, so I felt right at home.

The PR team, with my old buddy Ken and his event coordinators were based further into the complex of expensive holiday homes and I was taken for a look around by the lovely Lydia Hayley. Lydia and I had worked on many projects together and she was perfect in PR with great logistics and events coordination skills. Her previous employment in the Metropolitan Police Force made her a perfect fit. She didn't take any messing around and she called a spade a spade.

268

The mahooosive holiday home that we were using was also used by a very famous rock star and it was uber cool. The event tech guys had even rigged up underwater lighting in the pool with XF graphics at the bottom and what's not to like about that?

Inside the house they had also set up flat screens playing rolling videos of the XF being tested along with interviews from those who worked on the car. There were also areas set up with internet connections for the journalists so that they could ensure their reports were sent back to head office as soon as possible. When not working there was a games room with PlayStations and a pool table. I had never seen anything like it and I was so impressed.

For the evening dinner when they arrived there was another area of the complex dedicated to Jaguar as a private function room accompanied by a rather swish vodka type bar.
It all sounds very glamourous but as a journalist once told me, "first impressions always count."
The place was buzzing and I really enjoy that. Everybody has their own job to do and they just get on and do it.
Having checked and signed off the first batch of cars they head off to get valeted by our own on-site team of detailers. They work miracles and when they are done you could eat your dinner off them.

With some free time, Lydia asks me if I can team up with their contracted photographer Rob Till. Rob's job is to make sure we capture all the cars being prepared and then being used on the road for future press articles. They also want lots of detailed shots that we can give out to the press guys who don't have their own cameraman.
We take one of the cars and head off into Phoenix to start collecting pictures.

Rob is from South Africa and is very familiar with Jaguar, having done previous work for PR. Within ten minutes I knew we were going to get on like a house on fire and we did.

I was very keen on photography myself so at every opportunity I was asking questions and seeking advice on how to take the best pictures and Rob, bless him, went out of his way to show me everything. By the end of the trip he convinced me to buy a stunning camera, as a Christmas present for my wife, from a local store at half the price it was in the UK.

As part of the full Rob Till service he spent ages setting the camera up for me and all I had to do was wrap it up for Christmas.

We became very close during the trip and developed a great working relationship. During filming he would sit in the back, or out the sunroof or windows of the Land Rover whilst I drove the XF behind it. With the use of walkie talkies he would tell me to move right, move left, back off, speed up, pass us then brake and when the road was clear he would get me to drive within inches of his rear bumper to get close up shots. It was great fun and when you see the final images you realise just how good he was. He was also a fantastic wingman during the event and during the evenings at the bar where he shared many great stories with the team.

I have to say that he taught me everything he knows about photography and to this day I still know nothing (only kidding Rob).

With everything ready and the plane due to land at Sky Harbour Airport carrying some of the world's top journalists, it was show time.

A number of Jaguar Cars were ready and waiting with chauffeurs to take the guests to the complex. On arriving it was time for them to chill out and enjoy a relaxing evening to make sure they were fresh for the drive the next day.

In the morning our events team were in early and we selected several cars and had them all lined up and ready to go. Route books had been made so that they would never get lost and were placed on the passenger seats. Details of all the emergency comms were on a card for their wallets and all that was left to do was stand up and do a presentation about the new XF and the plan for the day.

Having done my pitch, my task for the event was to head off thirty minutes before the rest of the team and act as a scout for the route. At the head of the field it gave me the chance to report back if there were any issues, but to be honest we were so far into the desert there was never going to be any traffic issues but they do seem to close roads every now and again without warning so it's always best to be prepared (remember the German barrier now).

Another task of being the scout car was to get to the mid-morning coffee stop and get everything arranged for their arrival, so each day my passenger would be one of the girls from the PR team. It was great for me because I then had somebody to talk to for eight hours and they in turn got a day out to see the route. My first passenger for the week though was the head of PR, Ken McConomy.

With everything ready the train of cars would arrive with the guests for a well-earned break and some caffeine. Their arrival was our cue to depart for the next leg to the half-way lunch stop.
Our scout vehicle was the gorgeous XK sports car and we were loving it. The music was on and all we had to do was burn up the miles and smile for the next couple of hours.

Being good mates meant that we chatted about everything. We discussed the trip, the cars, the team and the guests for most of it and spent the rest of the time talking about Manchester United, any sort of gossip and food.
As we headed towards our lavish lunch stop, Ken suggested it might be better to get a meal inside us before we arrive, and in the town there just happened to be a KFC in front of us. Ken announced that it was the food of champions and we had to stop. So there we were, only a couple of miles away from the best restaurant in the area and we were sat with a mega bucket of southern fried chicken. We pigged out and it was delicious.
The ranch house restaurant was amazing, but we just accepted some water and made sure everything was ready for the teams who arrived only minutes after us.

They parked up all the XF's in a line in the scorching heat and headed off for some air-conditioned time out.

With clear blue skies the colours of the cars looked amazing in the sunshine and they make you feel very proud of what we had achieved to get to this point. I always feel very lucky too, because it takes thousands of engineers to develop a car and hardly any of them get to see this. For my team back in the XF Vehicle Office though its only right to send a lot of stunning pictures of the cars and the scenery, and let them know that it's a dirty job and all that, but somebody has to do it.

Having made sure all is ok with the cars and the guests, Ken and I set off for stage three but the Mega buckets of KFC are sitting heavy and instead of looking like two super hero's setting off for their next adventure we are more like Fatman and Dobin.

Our next run is to a place called Apache Junction where it all gets a bit more historical.

The trail was used as a stagecoach route that ran through the Superstition Mountains and was named after the Apache Indians obviously. The route runs past the Theodore Roosevelt Lake, the Tonto National Forest and onto Apache Junction. The original direct road is still unpaved, and we have to use the more modern Route 88 but we still visit all the scenic points. The trail was completed in 1902 and served as the main route for the construction of the Roosevelt Dam which would provide the water supply and hydroelectric power to Phoenix.

The area was stunning and you can just imagine what it must have been like after the civil war.

As a kid I loved playing Cowboys and Indians and would always watch the big cowboy movies with John Wayne or Clint Eastwood, and now here I was, following the same trail as they did. It would have been great if Ken and I were on the iconic Appaloosa horses but at least our current steed had a few hundred horsepower for the trek.

Its days like this that you have to pinch yourself really hard to make sure you are not dreaming. Driving along in our flagship sports car with one of your best mates through sun kissed canyons as the sun starts to set over the mountain range is just very special and we both know it. We look at each other and we are both thinking the same thing. Shall we call Steeley back in the UK to tell him just how cool this is? Oh, go on then!

With the sun casting really long shadows behind the giant cacti we roll back into town after a long day.
The team take the XK away to prep it for tomorrow and within thirty minutes our posse arrives with all the XFs.

I wander around the guests to see how they have enjoyed the day and the feedback is very good. They are impressed with the car and had great fun driving it on a fabulous route.
All the technicians take the cars away to the back of the tent and park them up. Their job the next day will be to clean and prep them all while we repeat the drive for the next batch of guests who will be driving the cars already prepared by the team. The attention to detail by the event guys is impressive and always noted by the guests.
By the time we have shut up shop it's about 9pm when we return to our hotel and its straight into the local restaurant for food and beer until none of us can keep our eyes open and we all retire for the evening.

After a quick shower I call home to say hello to Danielle and tell her what we have done for the day. I tell her that I miss her but it's always difficult to get that over on a trip like this but its true....honest!
I'm so knackered from a day at the wheel that I just strip off to my boxer shorts and pass out on the bed for a well-earned sleep.

I'm woken at about 1:30am by my phone ringing and it's Lydia. "Hey Dave, you couldn't just come back to the complex to get a car out for us could you? One of the journalists and Ian Callum want to see a specific interior colour."
I desperately want to go back to sleep but twenty minutes later I'm at the complex and opening the tent with Lydia, and as always, it's a case of Murphy's Law. The car we need is in the middle of the tent so I have to move a few cars to get it out. We

273

finally get the car and we set off up the hill to the hosting area and Lydia heads off to get them.

They both wander over to the car and I am prompted to open the door and turn on the interior mood lighting.

With it looking splendid they put their heads in, take a quick look and the journalist exclaims "that's flipping brilliant Ian." Without any other words they wander back to the party. Lydia is mortified beyond belief. "Dave I'm so sorry for getting you out of bed at this hour for that."

There is nothing to be said and I can only smile because if he writes a fantastic report about our new car then its job done!

The whole process is repeated for the entire week but with different passengers it makes it fun every day as you have different things to talk about and learn more about the people you work with. One of the best days was with Manjana Roth. She was of German descent but worked for JLR at Whitley. She could speak seven different languages and had a number a fascinating jobs around the world and was brilliant in her role. Her life was fascinating, and I really enjoyed talking with her for the day and I don't think we discussed cars, motorbikes or Man United once.

The event turned out to be a major success and the coverage in the motor magazines was amazing and well worth the effort. I even managed to get my face in the paper driving one of the cars with Ken Gibson from the Sun newspaper.

When it came out my parents were thrilled and couldn't wait to call me with the news.

After the last drive the whole team decided to push the boat out a bit which was led by my Senior Manager Andy Wyman. He is well known as 'hollow legs' as he seems able to drink so much with no effect and we figure that's where the drink must go. With the evening in full swing, Rob and I decide that we want to do some photography the next day at an amazing Air Force base near Phoenix and both Andy Whyman and our designer, Wayne Burgess agree to join us. We head off to bed and leave the professionals to it.

At 7am it's not surprising that there is just Rob and I in the car and we are a little surprised when Wayne arrives twenty minutes later. We call Andy but he admits defeat and goes back to bed.

With Wayne asleep in the back we head off to a place called Davis Monthan or as most in the know call it 'The Boneyard'. It's a resting place for all the ex and surplus military aircraft and there are hundreds if not thousands of them. When we arrive, we head over to the reception at the museum and manage to get hold of one of the managers and we give him the low down on our trip and the new XF. After letting him have a drive around the car park he allows us to take the car in for a quick photo session and Rob is chuffed to bits. Planes and cars always go well together and it makes for some great photos.

We place ourselves in an area with about fifty Hercules aircraft. It's an incredible sight.

The museum is our next stop and as we are going in we are talking to some of the guys from the hanger workshop who ask if we have come to see our old girl land. I question the bit about our old girl and find out that an old Shackleton originally from the UK is due to arrive at the museum any minute. It's a coincidence and suddenly we are made guests of honour to watch this fantastic moment. The old girl comes into view and does a few passes of the air base before finally landing for the last time in her life. Pima Air Museum will now be her final resting place.

To make it even more amazing is that the plane was donated by Air Atlantique in Coventry only a couple of miles away from Jaguar Whitley. Sometimes you are just in the right place at the right time!

We spend most of our down day at the base taking it all in and enjoy seeing so many aircraft. If you ever get a chance then go onto Google Earth and check it out. It will blow you away!

All that's left to do is take a relaxing drive back to Phoenix and start the whole process of packing up and heading out. The whole event was a major success for Jaguar that generated loads of column inches in all the magazines and put our new baby on the map.

# Monaco or Bust

It wasn't too long after Phoenix that we repeated the drive event in Europe but on a bigger scale. The early reveal in Phoenix was for our selected magazines and Jaguar Ambassadors but this event was for all the newspapers, smaller magazines and freelance journalists as well as staff from many dealerships around the world.

In principle the whole event worked just like Phoenix but on a far bigger scale with over fifty cars and a much bigger tent!

Unlike Phoenix though, our location was a little less glamourous as we were in an area of Nice that normally had open air markets. It wasn't pretty but it served its purpose.

I flew out to Nice with most of the Vehicle Ops team and arrived at the beach front hotel in gorgeous sunshine. For a brief five minutes I looked out across the sea and nearly convinced myself that I was on holiday. For most of engineering who don't get to go on these events they actually do think it's a holiday!

I'm removed from my daydream by Brian Pearson who beckons me to check in with him. Brian has been with Jaguar for many years and has spent much of it doing press car prep and events. During previous trips we have had a lot of laughs and I know it would continue on this one.

Having checked us all in we are all straight back into the fleet of Land Rovers and heading off to the centre of Nice and our new base. We open the doors to the tent to find all fifty cars parked up like sardines. There are tables and chairs for the engineers and supervisors and loads of stillage's with spare parts, books, stationary and everything else we might need to run a fleet of cars.

The boys have done all of this before and set off with military precision to arrange the tent into a working garage. By the end of the day we have the first batch of cars ready to go out on road test the next morning and all the parts are sorted into cupboards and racks.

Before we leave for the night, our security team arrive with the biggest and meanest looking Alsatian guard dog to watch over the tent for the evening. He looks like he will eat you but over the next couple of weeks I intend to become his friend.

Back at the hotel the fun and games begin in style. We decide to have a curry and arrange to meet in the reception at 7.30pm that leaves the team and hour and a half to go to the gym or relax in the pool before getting ready to go out.
Davey Jones and I decide that we are going to use this trip to get fit and head off to the gym with a few others. Brian however decides to relax and have a nice soak in the bath.

When I arrive back at my room I get a call from Brian in a mad panic and I fly up to his room.
As I get there his panic turns into full blown laughter. He had only put the bath on and then fell asleep allowing it to spill over and flood the bathroom. We had only been here eight hours and he had trashed the joint.
Having sorted out the mess we finally head off with the team to the local curry house. They take us to an extended table and then bring us a round of drinks with compliments of the house. It's a nice gesture but we have no idea why until we see Brian getting VIP treatment. They have got him a chair at the head of the table and all the waiters are calling him Sir. We are about to put them right, but Brian gives us the look to say "shut up." With most of the food now on the table the waiters retire out of the way and Brian announces that we are in for a good night because they actually think he is Sir Richard Branson. We couldn't control the laughter at all and we had a fantastic night with Sir Richard.

Leading up to the press arriving we tested all the cars and signed them off ready for duty.
For the first day they would use ten cars so that we could then detail the next ten cars for the second day. When they arrived back after the drive their cars would go back into the system for cleaning, ramp checks, test drives and a polish ready for another day.

Cycling the cars in batches of ten per day was the most effective way to utilise the vehicles and resource and it worked like clockwork.

The press would arrive at Nice Airport and we would make sure that there were chauffeur cars parked up and ready to take them into Monaco for the evening. After a reception and meal, they would then head into the casino for a while and hopefully into bed at a reasonable hour.

In the morning we had ferried over the cars from Nice and parked them in a row outside the harbour hotel.
They looked magnificent and sat well with the expensive hotels and multi-million-pound yachts.

Following breakfast, the Events team would then take the press out on an all-day drive stopping off at various places on route for scenic photo ops.
At the end of the day we would all go again into Monaco to collect the cars and return them to the tent for more cleaning and prepping.
Having returned on one of the days I got talking with the local security guard and told him how much I liked his guard dog who was really friendly. He said I shouldn't be fooled as he would take me down on command.
He then showed me all his training gear and I asked if I could have a go. Both he and most of the Ops team thought I was crazy, but I figured that it would be a great demonstration that would also make them laugh. He agreed to let me have a go and I got suited up looking like a dog's dinner.
The guard got me to fake an attack on him and then make a run for it through the tent. I obliged with his request and set off through the shop as fast as I could. Within about twenty yards I heard yelping and barking as this massive brute hurled himself at me. His jaws clamped on my arm and he pulled me to the ground like a rag doll.
It was a tight hold and he wasn't about to let go. The team thought it was hilarious but not one of them fancied a go. It was quite scary, but it didn't put me off having another go and I set off again with the dog taking me out again.

Once he was recalled and I had removed the suit he was just another cuddly dog, but the troops still kept their distance.

With the event in full swing I was once again joined by our photographer Rob Till and we were given the job of doing lots of dynamic photos from around the drive route that could be shared with the press. It was great catching up with Rob again and we had a blast around Monaco and the surrounding area. Some of his shots made it into the press and they were impressive!

Back at the hotel we were in for some more entertainment with our technician Davey Jones. I had arranged to meet him in the gym but as I was heading down he was just leaving looking very red faced. He couldn't stop laughing and told me that he had decided to go down early and get some time on the treadmill before anyone else arrived. He was jogging with his earphones on when his iPod fell off the machine and he instinctively bent down to pick it up. The machine just threw him off the end and it was game over. I was in hysterics and poor old Davey was so embarrassed he just wanted to run off and pretend it never happened. Unfortunately that can't happen on an overseas event with your work mates. I went straight to reception to get a picture off the CCTV camera and wrote a full report of the moment and printed it out for all the chaps to read the next day! The whole episode prompted the birth of our own news sheet 'The Nice Daily News'. It went down well and kept the troops entertained.

It only took us a couple of days to get into the swing of things and what I really wanted to do now was a lap of Monaco. As a petrol head and Formula 1 fan it was a case of 'Needs Must'.
At the end of another long day the team just wanted to chill out at the bar and get an early night, so I took the opportunity to grab the keys to a pool car and made my way out to Monaco under the cover of darkness.

It was a short hop from Nice and then down the winding roads to the main harbour. I parked up by the hotel the PR team were using and walked around the quayside. It was a warm night and a light breeze rolled off the water and gently kissed the very expensive yachts. I had dreamed about being at Monaco during an F1 Race weekend but never thought about being alone stood between the Nouvelle Chicane and Tabac. It was very quiet and all you could hear were the bells on some of the yachts and the water lapping the side of the harbour wall.

I head back to the pool car ready to have a go at a full lap. I would have loved an XK or and XFR but I had to settle for the Freelander pool car instead.

I set off from the hotel and went right, down to Tabac and sharp left into the swimming pool section.

The Freelander rolled around the corners and we made it through to La Rascasse. It's a very famous hairpin and very difficult for the F1 cars to get round but the Freelander does OK. I pull through Noughs and floor it down the start and finish straight. The noise levels from my trusty steed are a little muted so I have to make up my own F1 sounds and gear changes. In my mind I'm thinking about Senna, Mansell and Schumacher and in my running commentary I manage to pass two of them under the brakes into Sainte Devote. On the climb through Beau Rivage I sit under the wing of Senna waiting to pounce but my expected route through Casino is not as anticipated. The diversion took me around the Casino but put me back at Mirabeau and the race was back on. Throwing the Freelander around Fairmont hairpin and staying tight was going to give me the chance of a tow through the tunnel and take the revised finish line at hotel Jaguar. This was it, neck and neck out of the tunnel and who would brake first? Side by side downhill into Nouvelle, it was now or never. With no room left it's all on the brakes then off to flick it around the chicane and over the line. If only it was for real. I had flashbacks to so many races during the lap and some great memories.

Having slowed down by the hotel and with hardly any cars about it seems only right to do it all again, only this time in a classic race against Jackie Stewart and Graham Hill.

I'm sure many people wonder why we use locations like Monaco and Phoenix instead of Nuneaton. The town centre ring road circuit of Nuneaton might be a challenge, but it has no race heritage and lacks that bit of glamour. Nuneaton may offer the scenic backdrop of a slag heap called 'Mount Judd' but it's not quite the same as the hills of Monte Carlo or the Superstitious Mountains in Arizona. To be fair Nuneaton does have a lot of cowboys and its fair share of Indians but it's still not a patch on the Apache Trail with its cacti and red boulders.

At the end of the day it's about attention to detail and creating the right ambience to go with the car and having some gorgeous backdrops for all the photography. We also want to make our guests feel special and part of the family.

It's all well thought out and when you see the press articles and the photos it's all worth it.

Monaco was beautiful beyond all my expectations and I hope to go back for a Formula 1 race one day. It gave me many great memories but the one I will end with is my flight out of Nice Airport.

As I stood in line for security I was just gazing around and as I glanced down at my bags I noticed the most amazing pair of female boots just behind me. I just had to see who was attached to these boots and legs, but I couldn't make it too obvious. I stretched out a bit and rolled my neck around a few times as if I'm easing off some pre-flight tension and then did the half a turn to get a good look at a fine and very glamourous looking lady. As I smiled, I realised that the man holding her hand was F1 driver David Coulthard who lived and had a hotel in Monaco.

He gave me the look and some raised eyebrows and all I could say was "Morning Mr Coulthard" and quickly left to catch my plane.

Au revoir Monaco! Je reviendrai avec style un jour.

# 23. E-Type Run to Geneva

From 1958 to 1960 Jaguar was secretly working on a sports car that would rock the world. Following its glory at Le Mans with both the C-Type and then the D-Type Jaguar was on a crest of a wave and they needed a new cat.

Malcolm Sayer had been busy designing a gorgeous aerodynamic body and William Lyons with his engineering team were putting together a chassis and powertrain to match.
Realising the potential of their work a prototype 9600HP was built and secretly handed over to a number of trusted journalists from Autosport, Motorsport, The Telegraph and The Times to prepare a road test feature for their editions in the UK during the early part of 1961. The planned launch of the car was to be in Geneva in March, and it left very little time to get the car over to the event.

Bob Berry, who was one of the Jaguar Executives, decided that the best thing to do was drive the car direct to the Parc des Eaux-Vives rather than having it transported on a truck.
He arrived just twenty minutes before the big reveal much to the relief of Sir William Lyons. As the car was given a quick clean Sir William said "Bob, I thought you were never going to get here."

After the reveal the press could not get enough of the new sports car and there was such a high interest in test drives that Sir William sent a message back to Coventry telling Norman Dewis to 'drop everything' and get another car down urgently. One of the drop tops 77RW was selected and Norman drove through the night and arrived to great applause from the press.
The car was an instant success as a true sports car with world leading performance. The show car 9600HP went on to achieve 150mph at Jabbeke on the Belgian Autoroute where it rewrote all the rule books and a star was born. The Jaguar E-Type.

The story about the launch of E-Type is well known around the world and for me it was something I had read about and talked about with the late Norman Dewis. His drive to Geneva overnight appealed to my sense of adventure and I could relate to it with some of the work we had done in engineering over the years. I had the pleasure of doing some mad dashes to the Nurburgring with prototype XK8's and some XF's to Portugal for press drives. The tension of meeting a deadline combined with some spirited and enthusiastic driving was a real buzz.

During late 2010 I got a call from our Heritage manager Tony O'Keefe to see if I could help with a very special drive event to celebrate 50 years of the E-Type launch. He was hatching a plan to take 50 E-Types down to Geneva and recreate the moment in the Parc des Eaux-Vives and then head into the Geneva Motor show where the car I was working on the XKRS was making its debut. It was perfect and my boss Kev Riches has no hope of turning me down when I gave him the sad puppy eyes look. Luckily Kev has an E-Type and ended up taking it down to let some of the press repeat the drive route.

With lots of planning by Tony the event took shape and 50 customers were set to join us on an exciting tour and the schedule put together by Tony was very impressive as you can see.

### Sunday 27th February 2011

The event would start at the historic Coombe Abbey hotel and in the evening, participants would have the chance to relax and get to know each other at a wonderful Civic Reception at the Coventry Transport Museum, home of one of the UK's finest collections of motor vehicles.

### Monday 28th February 2011

The cars would depart from Coventry on the first leg of the route to the famous Castle Combe race circuit where there would be a chance to drive a few laps of the circuit before enjoying lunch. From Castle Combe the route would continue south over Salisbury Plain to Goodwood for a night at the Goodwood Park hotel and dinner in Goodwood House, home of Lord March.

**Tuesday 1ˢᵗ March 2011**

A great start to the day, driving up the Goodwood Hillclimb and then on to the Goodwood Circuit for a few laps. The route would then travel east across the beautiful South Downs to Folkestone to cross the Channel using Eurotunnel. From Calais the route would continue south to Reims, where there would be a Champagne reception and dinner at the famous Tattinger Champagne House.

**Wednesday 2ⁿᵈ March 2011**

The day would start with a lap of the historic Circuit de Gueux in Reims. Travelling at a more leisurely pace, the journey followed a scenic route south towards Dijon, stopping off at the Dijon-Prenois Circuit. Here there was an opportunity for light refreshments and laps of the circuit, before continuing to the beautiful Château de Chailly estate for dinner and overnight accommodation.

**Thursday 3ʳᵈ March 2011**

The final leg of the journey to Geneva crossed the spectacular Haut-Jura mountain range on some very dramatic roads. On arrival at Geneva, lunch was hosted at the Parc des Eaux-Vives - the location of the original Press Launch of the E-type. For the next two nights, participants were able to relax and enjoy accommodation in the luxurious 5-star hotel InterContinental. In the evening there was an opportunity to visit the Geneva Motor Show providing a chance to see the very latest products on the Jaguar stand.

**Friday 4ᵗʰ March 2011**

A change of transportation with a boat trip along Lake Geneva with lunch included. The finale of the event was a spectacular Gala Dinner at the hotel InterContinental. The dinner was attended by a host of Jaguar personalities representing both the company's long and distinguished history and its bright and exciting future.

**Saturday 5<sup>th</sup> March 2011**

Participants were free to return to the UK at leisure.

This was a class event and I was overjoyed at being invited. My engineering role was to look after any issues with the cars during the drive, along with several technicians, but Tony always liked me to mix with the customers and talk to them about modern day development and testing which they find really interesting. I enjoy this side of it and feel very comfortable talking to anybody. As a youngster though I was extremely shy and anything like this would horrify me. At school I was so shy that I refused to run the individual 100 meters because everybody would be watching me, but I would run the 4 x 100m relay because I was part of a team. I would never run the last leg though because that's when the race was won or lost.

I think my confidence grew throughout my apprenticeship when I hit the real world. During my time on the shop floor at Brett's Stampings there was a lot of banter and you had to stand up and fight your own corner. Someone else who made me conquer my nerves and shyness was my Dad. He was a managing director with Ladbroke's looking after several betting shops but also running a couple of dog racing circuits and a very successful credit betting division. During his years he met many celebrities and did loads of gala evenings and events and was a very good speaker. It was during this time that he took me along to many of these events and he always pushed me into to conversations to talk about my apprenticeship and then my work in engineering. The more times he did this the easier it got and when I realised that people were actually listening and enjoying what I had to say it became a pleasure.

The other thing that Tony liked was having a safe pair of hands to do some driving if required, and with some track time at

285

Castle Coombe, Goodwood and Dijon planned he knew he could ask me to step in.

All of the E-Types started in Coventry and it was a fantastic sight. The cars were all highly polished as were their owners all ready to start an epic adventure.

We had a number of Jaguar support cars as well as a couple of our own E-Types for the event, but I started out with a very nice XF for the run down to Castle Coombe. At the circuit the owners were allowed to do some hot laps of the track but given the age of the cars and most of the drivers it was a little more sedate. For those that fancied something a bit more spirited they could jump into the XF and have a passenger ride with me at the wheel. For those that did they really enjoyed it and so did I. There were a couple of E-Types flying around though, one of which was a lightweight race car and the other one had a very experienced set of hands at the wheel with ex Formula 1 driver Martin Brundle. He was in his own E-Type and one of the guests, but it was great talking to him and getting him to share his experiences with the customers.

Following our exploits at the track we had another little drive to Goodwood for the night and a rather splendid meal at Goodwood House. It was one of those dinners when you haven't got a clue which knife and fork to use for what and you had to wait to see which side the person next to you took their bread roll from. It's so much simpler on test and development trips when you stand in the queue for a Big Mac or a bag of chips!

In the morning we had a real treat for the customers with a drive up the famous Goodwood Hill. Luckily for me I have had the pleasure of doing this in anger at the Festival of Speed and it's a hoot! We line up all the cars and then wave them on with 15 second intervals and they are loving it. Lots of cheering and waving is going on along with some enthusiastic tooting of horns. It's all very civilized and nothing like 'The Festival of speed'.

With all the E-Types gone I just can't help myself and leave plenty of room to stretch the legs of the XF. Giving it full beans off the line I leave a nice set of rubber marks (sorry Lord March) and make my way up the hill. Entering the first right hander though there is a massive problem. On my previous runs at Goodwood there has been hay bales and marker points that you are familiar with that define your braking points and set up your apex's but for our run today it's deserted. Exiting the first set of curves I head up in front of Goodwood House, but I feel lost. There are no crowds, no stands and no bridge. It all feels very open and very wrong. I'm suddenly aware that this is not a racetrack at the moment and is just one of Lord March's private roads through the estate. I wake up a bit at Molecomb corner and I'm back on it a bit for the climb to the curves around the old flint wall. My enthusiasm is curbed once again as a number of chickens scatter across the road and I end up taking a more leisurely drive to the top.

It's a great way to kick off the day and put us in good spirits for the long drive to Reims in France.

We reach the Eurotunnel and I end up getting my hands dirty. Most of my work mates would argue that I only get dirty when I fall over but in this case, it really was hands on dirt. One of the E-Types was suffering badly, and the gentleman and his wife were at a bit of a loss. We traced the problems to a dirty fuel filter and a poor contact in the fuse box. Both were fixed quite quickly so we were very happy and the customers now had some new friends for the trip. Whilst we were doing the work I got talking to a lovely fella and his wife who had their E-Type for years. He was keen to let his wife relax a bit and asked if she could go in the XF. It wasn't a problem, so I switched into his E-Type and he said he wanted a break too, so I ended up in the driver's seat. The lady was looking forward to some more creature comforts and I got Paul Bridges who was one of the other engineers to do the chauffeuring.

Driving an E-Type is always a thrill and you are always very careful, but when it's a customer car there is a bit of added pressure.

We eventually head off to Reims and after an hour I'm quite relaxed wafting along and enjoying both the sights and the company. The old fella loves his car and tells me chapter and verse about its history and all the drives they have been on with her. I love the passion that people have with their cars. After a few more miles he tells me that it was always his plan to get the wife into the XF. I asked him if that was her favourite car, but his reply was hilarious. He said, "Oh no, it's just that she rattles on so much I wanted a bit of peace and quiet."

At the next stop point I checked with Paul to see how he was getting on and he was a bit too keen to swap cars.

I rejected the swap deal and went back to the E-Type. I told the gentleman that Paul and his wife were getting on like a house on fire, so we were Ok all the way to Reims.

Our arrival and evening at Reims was a bit special because I do rather like the taste of Champagne. I don't mix in the Champagne circles, but I do enjoy the odd bottle every now and again for special occasions obviously.

Tony on the other hand does mix in those circles and had arranged a fabulous evening with Tattinger. It was an excellent evening mixing with our customers, sharing some early stories of the event so far and drinking a few glasses of their finest champagne. There was another feast fit for a king and we were all presented with some rather fancy Tattinger cufflinks that I still wear for special events. It's fair to say that if development testing was like this all the time we would never have finished testing any of our cars!

During the course of the evening I got chatting with a really nice couple. They had been everywhere in their E-Type and loved it. The lady then went on to tell me that she was photographed on the bonnet of an E-Type during the early 1960's as a model. I was fascinated by this and Tony confirmed the story for me. When I got back to the UK I found the picture and have to say she was indeed a very classy lady.

Our next bit of fun was having a lap of the famous Reims race circuit where Jaguar had much success with the iconic D-Type. We drove down to the old pits which sits on the side of the current main road. Over the years we have contributed to the refurbishment of the pit lane and have our own Jaguar marked garages. It makes an excellent place to park up for great photos.

After a road legal lap, without breaking the speed limit (honest), we set off at a very relaxed pace towards Dijon.

For this leg of the drive I switch cars into one of the last V12 E-Types built. The car is owned by the heritage team and is a fine example of the series. I tuck into the end of the cavalcade and immediately relax into waftmatic mode.

The top is down, the sun is out, it's a bit chilly but the heater is on and we have a stunning drive to enjoy.

You just live for days like this. I must have done something right to get these gigs because most people can only dream about them. We pass through small towns and villages and its fabulous watching the heads turn to the sight of all these cars. There is something very special about the E-Type though. Whenever we are on our own and stop at traffic lights you notice that people in other cars are looking at you and smiling. The gorgeous aerodynamic lines of the car appeal to everyone and as Enzo Ferrari once famously said "It's the most beautiful car ever made."

The car also changes you when you drive it. I have spent hours driving very powerful sports cars that you just need to drive hard and fast but with the E-Type you relax and enjoy it more.

The list of famous celebrities to own an E-Type is long and distinguished. Frank Sinatra, Steve McQueen, George Best, Tony Curtis, Princess Grace, Brigitte Bardot and George Harrison to name but a few and you can see why they did. During my stint to Dijon the car made me feel like a celebrity so why would you rush it. I took my time and milked the moment!

At Dijon we arrived at the famous F1 circuit and I couldn't help myself. The previous few hours were in the past and I'm now back into race mode looking for a suitable stallion to take on the circuit. Tony suggests that I stay in the E-Type to see what I think so I offer the passenger seat out on a first come first served basis and we are off.

Being one of the last cars built and one of the collection I'm very aware of the implications of a mistake out on the circuit but I still give it the beans albeit with a hint of mechanical sympathy.

We have a blast and give the old girl a few thrills around some of the tricky corners. It pulls well on the straights and is not short of performance. We pass quite a few of the customers who wave and cheer us on enjoying the rare sight.

I could have played there all afternoon but unfortunately the logistics of keeping a group of fifty cars together meant we had to move on after a couple of laps each. Looking back now it probably saved me a lot of money because I was starting to fall in love with the car and additional laps may have sealed the deal on a purchase.

Following our night at the gorgeous Chateau Chailly we pressed on for the final leg to Geneva. Our route was going to take us over the Haut-Jura Mountains which sounded fantastic. All the cars were in good order and their pilots were ready for another day behind the wheel. I managed the first stint in a drop top, and it was just what you needed in the morning. The bright blue skies and abundance of fresh air was just the ticket to wake you up. There really is something magical about the air turbulence wafting around your head when you get up to speed. Combined with the performance and noise of the car it makes you feel alive. If it makes me feel like this now, then what must it have been like in the swinging 60's? No wonder George Best and Tony Curtis had one.

With the coffee stop in sight I'm trying hard to find good reasons not to get out of the car. Maybe I should find some key issues with the car that need me to stay with it? Unfortunately for me I have to swap to let somebody else have their moment in time, but the timing is perfect. On the next stint we are climbing the mountain pass and by the time we get to the top it's much colder and there is a lot of snow. When I pull up next to our convertible the two lads from the heritage team are sporting thick coats and wooly hats. It's so warm in the Land Rover with its heated seats and steering wheel and I give them both my best smirk and look for a suitable parking spot.

It's another surreal moment. A plethora of E-Types arrive at the top of a mountain ski resort in four foot of snow trying to park for a coffee. It creates a stir amongst the locals and soon people are skiing up to take a look.

After our chilly pit stop, we are doing our descent into Geneva and our splendid cavalcade that has stayed together for most of the trip merges into the daily chaos of a major city. I'm feeling ever so slightly smug once again because I now have the benefit of sat nav but unfortunately the majority of the E-Types have just a route map. In an attempt to gain some road order, I try to corral as many of the Jags as I can and get them to follow me through the traffic. I start off with about six cars but even after five minutes I have lost two. At the next set of lights I hop out to tell them all to stay close and then jump back in. As I'm waiting to turn left several E-Types go past to the right. Maybe they know something I don't? We go left and stick to the Sat Nav. Heading up the road to our destination we see another gaggle of Jaguars going the wrong way but at least they are waving and tooting horns. Geneva did not know what had hit them. The locals at the bus stops must have thought they were dreaming. They don't see an E-Type for ages and then suddenly they see about fifty of them going in all directions. It was pure comedy central and I wish we had some video of it.
It all turned out ok in the end and all our guests either by luck or judgement made it to the Parcs des Eaux Vives.

Our arrival at the Parcs stunning restaurant marks the E-Types 50th Anniversary of when it was unveiled by Sir William Lyons to the worlds press. It's a very special moment for all of us and to ensure it feels right we move all the non-E-Types out of the way.
I wander back up to the restaurant and we have a special treat for the team. In 1961 Norman Dewis drove a second car to Geneva at the request of Sir William and we had that very same car with us today. We moved the car to the very spot that they used for the original 1961 photo shoot. Even on its own it was a stunning moment but the icing on the cake was having the legend that was Norman Dewis, to pose with his car. It really

doesn't get much better than that. I remember thinking at the time that it was a special moment but looking back it was much more. Along with nearly a hundred guests we had travelled for four days to celebrate a car's 50th Birthday in the very same vehicles built fifty years ago. We had visited and driven around four iconic race circuits, stayed at some amazing hotels and finished up at the exact location that it all started with the very same car and its driver.

If Sir William was looking down on us I'm sure he would have been as proud as we were!

How could you top that? Well, we did and all I can say is that Tony O'Keeffe is well connected because at our hotel for the Gala dinner we had the original E-Type Fixed Head Coupe on display. Quite rightly it stole the show and that's what the drive was all about.

So, what about the Geneva show? On any other year if I had flown to the show there is no doubt I would have been raving about it. However, after our previous four days with what Enzo Ferrari called 'The most beautiful car ever built' it was all a bit of an anti-climax.

The return to Coventry was a little more direct and not quite as glamourous but it did only take me just over 11 hours which the late Norman Dewis would have approved of, and he didn't have Suzannah Mullin singing to him all the way back either!

# 24. Reims de L'Elegance

The Reims de L'Elegance event from the 25th to the 27th September 2009 was a new adventure for Jaguar Cars.
This was its third running for the demonstration of competition cars, motorcycles and static exhibitions. In essence it's a bit like the Goodwood Festival of Speed and was already one of the main events on the list for historical cars.

The event takes place around the old grand prix circuit of Reims-Gueux which is about 7km just west of Reims.
It was established in 1926 and by 1948 it was seeing its first Formula 1 cars. Racing around this very fast circuit continued until 1972 when it closed permanently.
During its glory years, Jaguar did very well at the circuit. In 1952 Jaguar finished 1st with Stirling Moss in the C-Type and then again in 1953. In 1954 it was the new Jaguar D-Type that took the win and again in 1956 with the legendary Duncan Hamilton behind the wheel. Jaguar D-Types also took 2nd, 3rd and 4th in the same year making it a clean sweep.

With the success of the XFR and my Bonneville project the heritage team were keen to use the car. Luckily for me it was a package deal and they offered me a trip out to look after the car and do some demo laps.
As part of the weekend we also offered drives to Steve Cropley and Ollie Marriage who are both well-known motoring journalists. I had met both of them before and was really looking forward to the event.

We were taking a few racing cars with us for the trip including the Le Mans winning XJR-9 (Silk Cut Car), D-Type, C-Type, Group 44 E-Type and my XFR Bonneville. It was an honour to be in such legendary company with four classic race cars that had won so many races. The Bonneville car had reached 225mph to become the fastest Jaguar production-based car, so it was worthy of a spot at the event. It wasn't the fastest though as the XJ9-R could do 245mph on the Mulsanne straight at Le Mans. This was the car that drew me into Jaguar. It was my poster car and I spent hours drawing pictures of it in my spare time. I had watched it win at Goodwood and again at the Le Mans Classic and when you hear the engine it makes the hairs on the back of your neck stick up.

Many times I had mentioned to Tony O'Keefe that if the chance came, I would love to be a nominated driver for it at special events. It was even mentioned that I should go to MIRA for a couple of lessons with it, but it never worked out.

With the cars all loaded on to the transporter we headed off to France via the Channel Tunnel. We parked up and Tony announced that as part of a reduced cost package deal we had agreed to put the cars on show for a while. An excellent idea, and time had been scheduled in to do it. The cars were removed one by one leaving the XJ9-R till last.

As it was being unstrapped, Richard Mason suggested that I fulfil my ambition to drive the car and get it off the truck. Even though it was just being lowered and we couldn't start it, I still treasure that moment today. Just sitting in the seat with my hands on the wheel with it rolling was amazing. It's such a fantastic piece of our racing heritage. Thousands of Brits went over to Le Mans to witness the glory and the celebrations were epic. There were Union Jacks everywhere and the Jaguar name was back. As I crossed the line and pulled up just outside of the truck there were just ten of us and a few terminal passengers watching. It looked like a calm, collected and professional dismount from the outside but inside that cockpit there was as much joy from me as there was from Andy Wallace when he won the race. Secretly there was a tear in my eye and I had to take a deep breath before I got out. There were no further celebrations for my efforts, no champagne and no garland, just Richard telling me to sort out the XF and get it moved into position. I was back in the land of the living with a smile like a Cheshire cat.

The car went down well and was enjoyed by many passengers on their way to France. It's always good PR and a great way to keep connected with the public.

With PR duties completed we trundled off to Reims for the weekend activities.

On arriving there is never any messing about. Cars are unloaded and we are straight into PR mode. Tony and Ken have arranged a pitch outside the main town hall to promote the weekend show and it works well.

It doesn't cost you anything and it lets you get engaged with the locals and a lot of the fellow participants.

I find promoting the brand an easy job. Who could not fall in love and want to talk about our cars? They are all so special and each one has a fantastic story to go with it.

We only stay a couple of hours and then head off to the Reims track to drop off the cars in the paddock. We set them up for the next day and then head to the hotel to get ready for the evening.

As always, Tony has a brilliant night arranged for us at a famous champagne house called Tattinger that has been organised by an ex colleague Valentina Macchi. Whilst I'm a Nuneaton lad I do have a taste for champagne. I have no idea where it comes from as my mum didn't drink and my dad was either a beer or red wine man. As mentioned previously I never mixed in the champagne circles either, so I guess I just like the taste. It goes down so well and I just can't sip it. Normally my first glass is quaffed and then I must settle down out of necessity. Our night out starts with a coach drive from the hotel for our guests and by the request of Ken McConomy, I'm seated next to Autocar's finest journalist Steve Cropley. He is a big Jaguar fan, but it turns out he also likes card tricks and as it happens I have a few up my sleeve to show him. During a period when I was house sitting for a couple of months, without a TV, I read and learnt to do several card tricks. I really got into it and even briefly joined a magic group that shared tricks and presentation skills. We had an absolute hoot on the coach and Steve loved it. By the time we reached Tattinger he was just shaking his head demanding to know how it was done. Like any good magician you say nothing. The cards have come in handy on many trips and are great way to break the ice.

The Tattinger tour and evening meal was stupendous and it was a delight talking to all the guests about our cars and our heritage. The champagne flow was endless and delicious and there was no way I was doing any card tricks on the way back!!

Our weekend at the circuit started with bright blue skies and gorgeous warm sunshine. You could not have wished for better. As I walked through the paddock, I noticed a gaggle of classic motorbikes from my era of racing. I wandered over and there were Yamahas, Hondas and Suzuki's from the 70s but the one that really stood out was the MV Augusta. It was the bike made famous by the legendary Giacomo Agostini. As I joined a group of onlookers the rider of the bike came past and took station next to his pride and joy. I couldn't believe my eyes. It was the great man himself kitted up in his original leathers. I was gobsmacked. As I got my camera out, he was then joined by another legend Phil Read. Phil was one of Britain's best riders with seven world titles and he was one of my biking heroes alongside Barry Sheene. I hadn't realised that these guys were here and it would be awesome to see them riding again. After a few pictures I left the gathering crowd and set off to find our cars at the far end of the paddock.

In our tent we had a great line up for the crowds to enjoy. We had a 1952 C-Type, a 1956 D-Type, the Group 44 racing E-Type, the 1988 Le Mans winning XJR-9 and my very own XFR Bonneville car. Each car had a slot in the tent and a placard above it with the details of the car and its driver. I glanced up above the XFR to see my very own name plate and yet again I was grinning like a Cheshire cat. It was nothing like a pit garage sign at Le Mans but what the heck, it was cool and I made sure I got my hands on it at the end of the event. It's still hanging up in my garage at home above the door.

The first drives in my car were reserved for our two journalists Steve Cropley and Ollie Marriage. They both took to the circuit for a few laps and when they returned, they were blown away by the car's performance.
It's one of those cars that just looks right and as an aerospace engineer once said, "If it looks right, it will fly right."

With the guests drives completed I was next up for the demonstration laps and I couldn't wait. I had my Jaguar race suit ready and was busting a gut to get out on the track.

As I sat in my car waiting to go out I was tucked up behind the XJR-9. The car that had brought me to Jaguar was sitting there waiting to go out on the track. I desperately wanted to be driving it, but it just wasn't my time.

We were all pushed on to the grid and took our positions. Our group of Jaguars were kept together with the C, D and E-Types at the front and the R9 and XFR at the back.

We had about ten minutes on the grid which gave me time to gaze over at the pit lane and imagine what it must have been like for Duncan Hamilton when he raced here and won in 1956 with our D-Type. It's a strange feeling being in the same spot some fifty-three years later and just a couple of cars ahead is that very car. Money can't buy moments like this and it's all too easy to forget just how lucky I am. I remember sitting there with a huge smile on my face telling myself to enjoy every second because it will all be over in a few laps.

The sound of engines starting brings me back to life and I put on my helmet and then secure my race harness.

I tap the start button and the XFR bursts into life and sounds fantastic. Looking ahead I can see the team getting the R9 ready to start. Rich Mason climbs in and after a couple of minutes it also bursts into life. I thought my car was loud but my God the R9 was on a different level. Even with my helmet on it was just epic. I could see most of the team and the crowd putting their hands over their ears. To say I was excited was an understatement. Noise is such a big factor and I remember reading that a professor said that "the sound of the racing cars help create a sense of speed and power. They sound powerful and fast, so they are powerful and fast." Well, our 7 litre V12 was certainly powerful and extremely fast and the noise told you that!

In our brief we were told to do a couple of warm up laps and then give it a bit extra for a couple of laps as that's what the fans wanted to see. Our early and very expensive race cars were being driven by the Heritage team so they wouldn't be pushing them too hard. Mechanical sympathy and respect always comes first. For Richard and I we were allowed to stretch the legs a bit more and to make sure we had enough room we backed off from the team.

Even in just two laps the adrenaline was pumping, and I was taking some deep breaths. The tyres seemed to be sticking ok and the car felt good. Because we had set it up for Bonneville it felt fantastic in a straight line, but it didn't go around the chicanes too well and I knew I would have to be careful. We followed the circuit around to the main straight where the spectators were gathered and on exiting the right hander I knew Richard was going to floor it. The R9 took off like a wailing banshee and I gave the XFR everything it had. With Richard holding off a bit I ended up sitting nicely in his slipstream getting a massive tow and it was bonkers. There I was in a car that I had developed at Whitley sitting in the wake turbulence of a Le Mans Winner and my dream car. The rhythmic muscular contractions in my pelvic regions, normally reserved for Mrs Moore, were about to take me to the higher levels of automotive orgasms.

As we were coming to the end of the straight we were just starting to brake into a temporary chicane when a new noise blew us both away. Without a rear-view mirror in my car I had no idea that a Jordan Formula 1 car had joined us for a threesome into the chicane. Under braking it blipped its way down the gearbox and zipped through the chicane before we had finished braking. Wow…. that's all….Wow! It all happened so fast and by the time I got my breath it had gone, and we were both at reduced speeds for the tighter section of the track. It was definitely a moment worthy of conversation at the bar later. My opening line would be "did anybody get overtaken by a Formula 1 car today whilst racing against a Group C Le Mans Car? Oh yes, that would be me then!"

On the final lap we grouped all the Jags together and managed to get a great shot of them on the main straight to finish the day off perfectly.

After a quick shower and a change of clothes I headed off to the hotel bar to grab an early cold one.

I ordered a beer and the gentleman next to me asked if it had been a good day. I turned to say, "well actually it was a brilliant day" but I realised that it was Phil Read. I was taken by surprise and was a little lost for words. Having had one of those most unforgettable days I am now at the bar with a Motorbike legend. We talk for quite a while about cars and bikes and he said that he had seen our demo lap from the pits and loved the R9. He mentioned that he was staying at the hotel on his own, so it seemed only right to buy him another beer and introduce him to the team.

Unfortunately, the team didn't recognise him at first but when I introduced him by name there were a few raised eyebrows. I sat him down next to Ken McConomy and explained that Phil was an ex motorbike racer and within a few seconds Ken gave us a classic line, "So did you do any good Phil?" I had already started to snigger when Phil responded with "yeah I did ok. I won seven World Championships and eight Isle of Man TT Races." Ken nodded in appreciation and said "Oh, you were good then."

We had a couple more beers with Phil and had a good laugh too before heading out to eat. He didn't want to join us for food and said he preferred to grab another beer and retire for the evening.

All in all, it was on the list as one of the best events in my days with Jaguar. I managed to drive on a famous Grand Prix Circuit where a Jaguar D-Type won the 1956 twelve hour race, shared the tarmac with some of the most iconic race cars, been overtaken by a Jordan Formula 1 car and shared a beer or two with a motorbike legend. You just couldn't dream it up.

# Part 5 – Going the Extra Mile

Going the extra mile is about some of the things I have been involved with that are away from the core business but as a result of being with Jaguar

The link between Jaguar Cars and Jaguar Aircraft and a chance meeting in Scotland, realised my dreams to fly a military aircraft as well as establish some great friendships and a best man for my wedding day.

With our links and contacts it also helped us with a plan to do a custom paint finish on an ageing Jaguar Jet to celebrate its many years in service, and with one of those great Jaguar Pilots making it into the Red Arrows, I also manage to fly a Hawk Jet and assist them with some STEM projects.

# 25. Jaguar Car v Jaguar Jet

How many times have you sat at the traffic lights itching to outrun that sneaky little hatchback, who just raced up alongside you trying to take pole position? You take a few sideways glances to assess the driver and the potential of his chariot before making a snap decision to either saunter away from the lights without a care in the world or let loose every single bit of horse power you have to tear away from the lights like a scalded cat. If it's the latter then you will understand the feeling and buzz you get over the next ten to fifteen seconds as you edge in front of your challenger and credit yourself with a win. Doing it at Santa Pod raceway with equally matched race cars is even more of an adrenaline rush as you reach three figures over the quarter mile.

Since the invention of the motor car owners have been obsessed with speed and I'm definitely one of them, but where did this obsession begin?

The whole drag racing scene came from the dry lakes in the deserts of America during the 1930's. It became a popular activity amongst younger drivers and illegal races often went underground. They took place in all sorts of places but mainly used disused airfields just after the war. The first official drag race was held at the Goleta Air Base in California in 1949. The simple races pitted two cars against each other over the quarter mile and first one to the line wins. It caught on all over the world and now has its own governing body and organisation with thousands of drivers taking part to enjoy the thrills of speed and competition.

My chance in a big race was also going to happen on an airbase but this one was definitely not disused. The location was RAF Coltishall in Norfolk. It was the home to Douglas Bader in 1940 with No 242 Squadron with his Hurricanes but was now the home to the Jaguar Squadrons 6, 16 and 41.

I had been taking part in several events with 6 squadron as a result of my growing friendship with Sgt Mark Ashfield. We had met and became friends at RAF Lossiemouth in Scotland, and this continued when he was moved to Coltishall to join 6 squadron. Mark had a passion for cars and was very loyal to the Jaguar brand and its connection to the Jaguar Jet. The links went back many years and it's understood that Sir William Lyons had to approve the use of the name 'Jaguar' when Sepecat began production of the plane in the late 1960's.

Mark had contacted me regarding an RAF and families air show at Coltishall and asked if I could put a couple of cars on display in the hanger. I always loved these events as it allowed me to talk about our great cars, the testing and the fun we had developing them. It also allowed me to get up close and personal with military jets. Mark always went out of his way to ensure that I was well looked after, and I spent many hours climbing in and out of any plane that was static in the hanger. I even got some lessons in the flight simulators which eventually came in very handy!

For this event I had managed to get the XK8 from the James Bond film 'Die Another Day' and one of the Jaguar Formula 1 cars. They were a major hit over the weekend and the RAF guys were thrilled to have them. I ended up spending hours letting all the kids sit in the James Bond car and have their photos taken as well as running an impromptu raffle for a ride in the car at the end of the event. The major highlight though, to kick off the weekend's activities, was going to be a drag race between my XKR and the RAF's Jaguar GR4. The idea was hatched during the build up to the weekend and was approved by the top brass.

It wasn't the first time that we had done this though. I spent some time researching and dug out some pictures from the Jaguar archives of a race between an E-Type and a Jaguar Jet from the early 1970's that Jaguar Cars used for publicity. My race was just the curtain raiser to the event to entertain the crowd.

During the planning I was asked to sit with the pilot and the control staff to run through the race and go through the plan. The attention to detail was amazing and really impressed me.

Even though I was a civilian in a very fast car they treated me like another pilot who was planning his next mission.

We went through every scenario that could happen from taking the plane and the car out of the hanger, to waiting on the taxi way and through to the runway and the race.
We understood what would happen if the car had a puncture during the race, how I would react, which direction I would aim to go and what action the pilot and the plane would take. We even discussed what action to take should the plane have an engine failure whilst near the car or a bird strike during the run. I listened to every word and contributed to the discussions and risk analysis like a seasoned RAF pilot. I was hooked. Without a doubt I had missed my true vocation in life.

On the day of the event I was called into the crew room and advised that there would be a pre-flight brief at 11:45 for all of those involved with the race. Just how cool was that. I was living the 'Top Gun' dream and I wasn't even going to take off. I went back to the hanger to meet all the early guests and talk about cars, but my mind was firmly fixed on the clock and 11:45 could not come soon enough. At 11:30 I took a walk outside just to check the weather and I noticed that the crowds were far bigger than I was expecting, and they were all heading towards the sectioned off viewing area parallel to the runway. Well that added a little bit of extra tension!!!

I walked over to the crew room and made my way through to flight ops and took a seat at the back, only to be told to move to the front and sit next to Toby who was the pilot for the run. At precisely 11:45 the Squadron Leader checked his watch and formally commenced the brief. He welcomed all those present and then ran through the details for the run. I was assigned my own call sign, given a handheld comms radio and introduced to my co-pilot/navigator for the day Wing Commander Bob Judson. From that point onwards anybody watching or listening to the brief would have been totally unaware that I was driving a car. For every step of the way I was effectively a second aircraft that was going to line up on the runway for a dual take off. For me this was just magical and even I had forgotten I was driving an XKR. I was in the zone and you might have just called me and my passenger Maverick and Goose. Unfortunately, even with all the great planning I didn't even think of bringing the Top Gun soundtrack to play on the CD. Big fail or what!!

The brief first detailed how one of the Jaguars from 41 squadron was going to take off and be airborne for the spectacle armed with a photo recce pod to film the proceedings. Shortly after Flight Lieutenant Toby Craig and his new wingman David 'Maverick' Moore would taxi out from 6 Squadron holding area taking the left taxi way and then hold at the south end and await confirmation to use runway 04 North. On confirmation Jaguar GR4A would taxi out and hold the left-hand station of the runway and XKR would taxi to the right hand station. The control tower would then commence a countdown to launch and both Jaguar's would then be released for action. At the finish line my XKR would be pulled up to the far right of the runway and hold until the aircraft had departed the runway. On confirmation of safe exit, I would be advised to spin the car around and take station centre runway and hold until Toby returned at low level for a photographic opportunity. When completed I would then exit the runway immediately via the north exit and return to the 6-squadron hanger. Even as I write this, I'm reliving the moment and the hairs on my arms and neck are sticking up.

At our scheduled time slot, we left flight ops and strode out to our machines. My XKR was parked up alongside the flight line of Jaguar Aircraft and as Toby climbed into his cramped cockpit and martin baker ejection seat Bob and I slid into the far more comfortable leather semi bucket seats of the XKR. We established comms between the tower and Toby and I listened with excitement as the GR4 went through its checks for engine start. With the windows down and barely twenty feet away from the twin Ardour engines the start-up noise was epic. The whole sequence is just fantastic. First of all, you hear the electric motors spinning up the main shaft which creates airflow through the compressor and the combustion chamber. At this point you can hear the distinctive tones of the fuel pumps throwing hundreds of litres of avgas into the combustion chamber before an igniter makes the jet engine explode into life.

With the blades spinning the throttles are raised and more fuel is forced into the chamber until the engine is spinning at operational speed. Most of the initial noise comes from the compression of air and its flow. As it joins the fuel in the combustion chamber you hear deep booms. After combustion the exhaust gases are accelerated out the back of the engine and become the prominent noise that we all enjoy. It's the best noise on the planet by far and when you are this close you can feel the air pressure change. There is also a resonance and low frequency noise from the engines that you feel right in the pit of your stomach and it feels great.

With the GR4 nicely warmed up I start the XKR. With all the jet noise I can't hear the car and I have to check the rpm gauge to make sure it's running.
The control tower gives us clearance to taxi and Toby edges the fighter jet out and follows the yellow centre line to the left taxi way. Once he is clear I follow in formation keeping some distance from the heat of the exhaust. With the windows down we are getting the full experience of hot exhaust and burnt avgas. I'm sure doctors and specialists would highly recommend that this should be avoided at all costs, but you have to have some fun in life and this is right up there!

We approach the holding area and Toby moves the jet to make sure I'm not trapped by his exhaust wash. As he points the tail away the noise levels drop a bit and I'm able to talk to Bob for a minute. He takes me through the checks Toby is doing before he confirms to the tower that he is ready to go. The radio comes to life and I hear Toby confirming that his Jet is ready to go and requests clearance to enter the runway. The tower confirm clearance and ask us to proceed as per the brief. I watch the jet manoeuvre into place, and I drive the XKR around the back to my allotted position on the runway.

The tension and excitement is rising with every second and my heart is pounding like crazy. Bob is busy getting the video camera working and I have a quick look at all my gauges just to make sure all is ok. No engine warning lights or oil pressure lights are my only things to look for and all is ok. As I take a deep breath one last thought flashes through my mind. What if this all goes tits up and we collide and end up in a raging fireball down the runway? I have a quick shiver moment and shake off the thought as quickly as possible knowing that my plan of action should anything go wrong is to turn right and get out of the way as soon as possible and let the jet take off.

The radio is back on and the Tower has confirmed we are good to go and begin the countdown.

During the brief we agreed that to make the contest close Toby would start off without afterburners engaged so that we crossed the line at the same time just as he was going airborne. However, my actions in the next couple of seconds changed the rules of engagement.

As the tower counted down from ten Bob and I looked over at Toby who had also looked over to us for the final nod of approval. As the countdown hit five Maverick and Goose decided that the best thing to do was give Toby the 'Bird' in true Top Gun style. As the tower hit Go Go Go, I launched the XKR off the line as fast as it could go and we were both still giving Toby the 'Bird'. My lightweight XKR was always going to out drag the fifteen tonne plane off the line and in a few seconds we were well ahead and storming to victory. I looked in the rear-view mirror to see the plane getting smaller as we pulled away. I commented to Goose that this was in the bag, but he was not so sure and told me to check the mirrors again. As I looked up the 15 tonnes of military hardware was looming larger than life in my mirrors and the finish line was closing fast. With only a couple of seconds left there was an almighty noise followed by a colossal amount of jet wash turbulence as Toby left the ground and pointed his nose to the sky. The gorgeous sight of the twin Ardour afterburners in full flow confirmed that Toby had decided to do a full military take off as a direct result of our cheeky finger gesture at the start.

As he powered his way up, I could feel the full force of the engines on the car and for a second I had to hold on tight to keep it in a straight line. In all the excitement and with jet induced adrenaline I turned to Bob and exclaimed "That was f*cking EPIC."

With the jet out of sight the tower confirmed that we could turn the car around. As I looked over it was the first time I had noticed the rather large crowd. As we were there to please it seemed only fitting to do a full power slide and then bring the car to a stop on the centre line of the runway.

We sat there in a brief moment of silence wearing the biggest smiles you have ever seen as the RAF photographer ran out in front of the car. He lifted his camera and pointed it towards the car just as Toby returned low and fast down the runway. He absolutely nailed it over the car which shook us to bits, and we burst out laughing as we both ducked out of the way.

With the plane going ballistic we got the call to exit stage right and we raced out of the way so that Toby could do some pilot stuff for the crowds.

On our approach to the aircraft pan a lot of the guests had formed a nice ring around the jet parking area so with the Wing Commanders blessing it was time to entertain the troops with a few doughnuts and a lot of smoke.

We parked up and got out to a great round of applause, so it obviously went down well. Everyone was cheering apart from one poor fella in his squadron overalls who was dispatched by Bob to ensure that all my bits of tyre rubber were removed from the concrete ASAP just in case they were sucked up into the returning jet engines.

We strolled back to the crew room and were joined by most of the squadron and finally by Toby where we both gave our accounts of our duel over a few cold beers.

It was a very special day for all concerned and by the amount of happy smiling faces from all the guests and families that stopped to see us afterwards it was enjoyed by all.

I'm sure you might ask, well what did you get out of it? Well, from a sales point of view we were not there to sell or promote our cars. We were there to support the RAF and the squadron and add an extra bit of interest for all the families that stand by and support their loved ones who serve our country and protect our skies.

Hopefully many of the younger ones will remember that day when they touched or sat in the James Bond car, had their picture taken next to the Formula 1 car or watched an XKR flying down the runway. And that's how brand loyalty begins.
As for me......well I got to be Maverick for the day and went home smelling of burnt avgas. Does it get any better than that?

So, who won the race?  Well a Jaguar obviously!

# 26. Piloting the Fastest Jaguar

In my final year at school I decided that what I really wanted to do was be a military pilot in the all new Tornado, but on finding out my vision was not up to the 20/20 standard I was encouraged to look at possibly being a navigator. Being a back seater was not on my agenda. The alternative option of flying prop planes didn't appeal to me either. So with a deep sigh and a heavy heart I ended up looking at apprenticeships and eventually found my way into commercial/aerospace engineering. My first job and a reasonable wage allowed me to pursue a private pilot's licence doing a number of lessons in a Cessna 152, but it just wasn't a fast jet and I had to face the facts that it wasn't meant to be.

In joining Jaguar I never thought it would lead me in the direction to fulfilling the ambition of my younger years to fly a fast jet.

With the launch of the XK8 Paul Freeman and I were invited to attend an air show at RAF Lossiemouth in the North of Scotland, and we were asked if we could take a couple of cars to put on show.

We managed to get ourselves a coupe and convertible and drove up to meet our contact Sgt Mark Ashfield and his wife Karen, who had organised our hotel near Elgin and the itinerary for our visit.

The RAF day was for friends and family of 16 squadron who flew Jaguars, and obviously they wanted to form some links between the plane and the car, and we were more than happy to oblige.

We managed to get both cars on static display with a jet and I was in my element. With the aerobatic display Jaguar doing its thing I found myself intoxicated by the smells of avgas and the noise of the Ardour engines on full burn. Whilst watching the display I tried to imagine what it must be like with all that power pulling through 'High-G' turns, but it was going to be something else with far less power that would take me to the skies first.

Unbeknown to me Mark had arranged a little surprise with a flight in a Tiger Moth with chap who was the spitting image of Lord Flasheart from Blackadder goes 4th. We rolled up to this old barn hanger just as they were pushing out the old De Havilland DH82 (Tiger Moth). It was bright yellow and looked stunning.

My pilot for the flight came over and introduced himself and he even sounded like Flash!
I had a flashback to that awesome sketch in Blackadder when Lord Flasheart advises George to treat his kite like he treats a woman, and I begin to snigger to myself.

I was strapped in with my full flight suit, leather flying cap and goggles. Flash was ready behind me and in no time at all the late 1930's engine was firing on all cylinders (for the moment anyway).

After the slapping of thighs like old airmen do and round of Tally Ho and Chocks away we were bobbling down the grass runway. It's not quite the pace of the Jaguar on full burn or even your average car under mild acceleration but after what seemed like an age and with the benefit of going downhill the old kite lifted into the air and with gentle pressure on the stick it climbed gracefully into the sky. As we levelled off, I could feel the heat from the engine over my face and also breathe in the oily smell of the exhaust. The views were amazing, and we were just wafting around at a very sedate pace and it was lovely but then Flash bellowed from the rear about doing some air show thing. I nodded and gave him the thumbs up thinking he was just talking about the RAF day but what I didn't realise was he was telling me we were going to be part of the air show. The next minute he flipped the plane on its side and flew the length of the runway before going into a steep climb with a wing over into a return dive. I was so busy checking the structure of the aircraft thinking it might just break up that I missed the view of the spectators on the ground.

Flash was loving it and told me to hold on tight we're going for a loop. With full throttle he pulled it up and over and then suddenly there was a massive bang and clatter and my entire life flashed before my eyes. Dear Lord what am I doing in this old kite with Lord Flasheart at the helm and I'm not even wearing a parachute…. I'm doomed!

From the noise it was clear that we were definitely NOT firing on all cylinders anymore that's for sure, but Flash had it all under control and brought it back down to earth a little quicker than we took off.
It was still a very cool thing to do and I loved every second of it and it was so good of Mark to set it all up for us.

The weekend was a major success and it also established a friendship that continues today with both Mark and the RAF.
Mark moved down to Coltishall in Norfolk to join Jaguar 6 Squadron and was keen to continue our personal and working relationship and it worked out so well. We did RAF and Family day air shows and made sure the link between car and aircraft was always there. If the RAF had any event that required our support I was always keen to help out and I knew it was really appreciated but I didn't realise just how much until I was called in to see Mike Sears in the Officers Mess on one of the event days I was helping out at. He sat me down and said he was made aware that I wanted to join the RAF and fly fast jets and that he would like to make that dream come true. I sat there stunned for a minute or so just taking in what he had just said. With a broad smile he confirmed that yes, we were going to fly in a Jaguar Jet, and I had two choices. The first was that I could go up next week for a quick flight around or if I could wait a bit and maybe do some simulator time, I could join him on a training sortie. No brainer really, I will wait!

So, the date was set for the 21$^{st}$ June 2002 and it could not come around soon enough. On a few visits to Coltishall they managed to get me some time in the simulator and I also went in to get checked out in the ejection seat and understand all the things that you need to know if you have to eject.

When the date arrived, I was ready!!

311

As it happened it was also the day that England went on to play Brazil in the 2002 World Cup in the quarter final and when I arrived at the base all the squadron were watching the TV. Apparently, they had come in early to get the jets ready so that they could watch the match. I met up with Mike Sears and joined the team to watch the start of the game. I'm a big football fan but on that day my mind was elsewhere but I did manage to jump up and celebrate when Michael Owen put us 1-0 up. Not long after the goal I was called in to see the flight doc for last minute health checks and a pressure check of my ears. I had to hold my nose and then blow hard to make my ears pop and if you couldn't do it then you were not going flying which was something to do with pressurisation.

I passed all the checks and was taken down to the crew room to get suited and booted. I had to wear thermal trousers, top and thick socks and then my flight suit, some flight boots and my all-important G-suit. Putting this lot on was a good sign that we are going to be going fast and pulling some high-speed turns.

I waddled up to the next room and was fitted out for a flight helmet and they also talked me through the communications system, oxygen system and how to move the visors up and down.

I was fully kitted up and had a look in the crew room mirror to make sure everything was connected but I was just looking and thinking YES that's me in there and YES this is so cool.

Mike collected me from the locker room with the bad news that Brazil had scored to make it 1-1. Ordinarily I would have been gutted with this news but just for today I hoped that England might just get a winner and cap off a perfect day.

We picked up our wingman for the trip Graham Duff and headed out from the ops centre onto the flight pan where our two jets were sitting looking splendid in the early morning light. Mark was already out by my jet waiting to grab some pics and to get me strapped in. It was a nice touch because he was the one who had made this day work for me and for that I am eternally grateful.

I climbed into my Martin Baker ejection seat hoping beyond hopes that I would not need to use it but also safe in the knowledge that I at least knew what to do if I did. Up until then there was a lot of banter but the moment they started strapping me in it was all taken very seriously. The main harness straps were done up good and tight and then the legs straps were connected that would pull your legs into the seat if you needed to eject. The comms and oxygen lines were then connected and I was able to start chatting with Mike up front. As he started doing the flight checks I was able to follow him through it to make sure all my gauges and switches were also working, and the safety pins and catches were all removed and disengaged.

Having completed the checks Mark leant back into the canopy to wish me luck and enjoy my flight before shutting and latching the canopy down. If you are at all claustrophobic this is now the point you would panic as you are sealed into your very tight space bubble with a rubber mask over your face and not much room to move. It's a struggle to breath but Mike quickly informs me that you really must breathe in hard to get the oxygen through and when it does it feels great.

The engines are now up and running and it's getting very real. I watch the rpm gauges rising and watch in amazement as the whole instrument panel starts to light up. I locate the critical gauges, altimeter, fuel, airspeed, artificial horizon and confirm all is ok. The head up display is activated, and I confirm that a number of other switches are also activated as Mike walks me through the rest of the process. Luckily the simulator practice worked because I start to feel comfortable with the aircraft and begin to relax a little bit.

Ground control confirmed our exit from the pan, and we start rolling out to the taxiway. I catch sight of Mark and give him a wave and we are now rolling to the left of the airfield to a holding point just short of the runaway. Mike tells me to look out of the cockpit and wave to the plane spotters in the layby who are taking photos as they do every day. I used to do the same thing but now here I am on the other side of the fence in a multi-million-pound military jet.

Ground control give us permission to enter the runway and both jets line up together for a joint take off. In what seems like an age the jets are spooled up ready for take-off and then we finally get the clear to go.

RPM rises and the jet starts to lift before the brakes are released and we are on our way. The initial acceleration is not as good as any of our sports cars, but it is a bit heavier at about 11000kg.

As we hit our take off speed we rotate and start to climb and then you feel the power coming in and the speed rises quickly. On the comms Mike gets me working quickly and instructs me that I have control of the aircraft....BOOM....WTF.....that was quick. What a moment to go from passenger to pilot. There was no time to get ready or think about what you were doing and maybe that was a good thing because you don't have any time to panic, that is until you look out of the cockpit to see a another very large Jaguar jet right next to you and I guess the natural reaction is to move away because that's just what I did and Mike was soon on comms to tell me to keep it straight and level and don't even think about our wingman because he will follow us. It's the most unnatural thing to do and I can't stop looking out at the gap between us. When you fly on commercial planes you sit and look out of the window and occasionally get to see other aircraft passing below or alongside but they are all well out of the way and you have nothing to worry about but this is just unreal. As Mike mentioned he will follow our every move and I notice this when I make adjustments with the control stick. How on earth the Red Arrows do this is beyond me!

Having made our way to the first checkpoint Mike instructs me to make a turn and head out towards the sea and an area called 'The Wash' where he resumes control of the aircraft.

Within this area there are some targets on a sandbank including a warship and Mike informs me that the frigate is our intended target. Both jets fly past and pull up to then circle around for our first pass. Graham pulls in behind us now and we lead the way to target. Our first pass is a simple nose to target and release allowing the small 15kg practice bomb to glide its way onto the hull of the ship. As we pull away Mike lights it up and pulls us around for our next run. During the turn we pull about 4g and it really gets you. As I had been briefed beforehand you have to strain a bit before the turn is made so that you have control of your diaphragm before the G-suit inflates.

The next run has us coming in fast and low before pulling up sharply to toss the bomb to target and our exit is a rapid climb to a safe altitude. I'm starting to feel it even more now but after a good lung full of oxygen I'm a bit perkier.

As we move around for the next run Mike asks if I have seen any of the impacts but unfortunately, I haven't.

The next run changed that though as we skimmed in for a fast release and the exit was completed inverted so that I could see the bomb as it impacted on the frigate.

Mike flipped the aircraft into level flight and without any warning pulled it hard right into a High-G turn before I had a chance to prepare. The G-suit clamped around me keeping all the blood locked in the upper half of my body, but it did it before I had the chance to strain. It was at this point I knew I was going to throw up. On comms I let Mike know and he kept the aircraft locked into the turn until I had switched off my radio, unclipped my mask and got the sick back in place. Once he got a thumbs up from me in his rear-view mirror he pulled out of the turn and the G-suit deflated which in turn allowed the contents of my stomach to propel itself into the sick bag. Once the bag had been dealt with and stored, I was back on comms and oxygen. Mike was laughing but advised me to take control of the aircraft and take in more oxygen as it would make me feel better and he wasn't wrong. Apparently, the request to make me sick came from Mark who I thought was my mate.

Mike gave me some new co-ordinates and he allowed me to navigate my way over to RAF Holbeach where I became No 2 to Graham and followed him into the live firing range where I got the chance to watch 32mm shells streaming down onto target from both his gun barrels and it was very impressive.

Having used up all the ordnance we pulled up into the blue and cloudy skies and had some fun making turns around the clouds which I have to say was at my request.

As we were now out over the North Sea Mike asked if I would like to do some dog fighting and gave me the option of being chased or do the chasing. Well, what can I say, as a combat veteran on my PlayStation I was sure I could chase Duffy and take him out but as it turned out reality is much harder than Virtual Reality and Duffy was gone from my sights in seconds. He pulled up alongside us and Mike asked if I would like to be chased and even though I put in a couple of nice manoeuvres Duffy was stuck to me like glue and called in a shot for kill.

Both Mike and Graham were slick and I was so impressed with their skills which is a credit to the RAF.

With fuel burning up it was time to return to Coltishall via Cromer again at my request. When I was younger, we had holidays on the eastern coast and I remembered not liking Cromer too much so I just wanted to do a simulated run on the pier with it lined up in my gun sights, childish but rewarding!!

The flight back in went far too quickly and I handed control of the aircraft back to Mike for a perfect landing. We pulled into the pan and shut it all down but before I could take a breath a certain Mark Ashfield was at the canopy grinning like a cat …

"pass me your sick bag then you wuss"

I climbed down from the aircraft, gave her a quick pat on the fuselage and then shook Mike Sears hand for making my dreams come true.

Graham came over to shake my hand and say well done and we headed off for fish n chips in the officers mess.

So, not only did I fulfil my boyhood dreams I also managed to be in the fastest Jaguar on the planet.

I have to say a massive thank you to Mark for setting it all up and making it happen, to Duffy for being my wingman for the day and also to the late Mike Sears for being my instructor for the day and allowing me to pilot one of his planes.

As I look at my photo's I can't help but think that you left this planet too early, but I will remember you every year on the 21st June when the sun sets on the longest day of the year.

Group Captain Mike Sears MBE
15 Aug 1966 – 24 Aug 2005 aged 39 – RIP

# 27. Let's Paint a Plane

Have you ever attempted to respray your bike or taken on the massive task of rubbing back your car and doing a full respray? It takes quite some time and there is a certain amount of stress that goes with it. So, try to imagine what I was thinking when the RAF called me to see if we could help them out painting a Jaguar plane. It sounded bonkers, but as always, we were well up for a challenge.

With a lot of military cuts, the Jaguar squadrons were coming to an end so that they could take delivery of the new Typhoon. Having made a lot of connections with 6 squadron over the years they decided to give me a call when they had an idea to paint the last Jaguar as a celebration of its long service history.

The squadron was the longest continually operated in the world with a total of 93 years from 1914.
As they had virtually no spare cash to fund the project they were looking for a lot of help and I was sure we could play a significant part.

First of all, I gave Joe Buck a call in the Whitley design studio as he had supported me with many hair brained ideas over the years. Joe was a motorbike and plane enthusiast just like myself and we both had a passion for military hardware. My brief to Joe was to come up with something special that would have a wow factor for the outgoing squadrons. Leaving Joe to sharpen his crayons I contacted another old mate in the paint shop called Gary Maltby. Gary had been with Jaguar for many years and was the leader of the paint shop at Whitley who were responsible for painting all our concept cars. I asked him if he would like to join us on this crazy project and he jumped at the chance. It's not every day that you get taken out of your comfort zone to do something completely off the wall.

After a week or so discussing various schemes with Alan Vernon from the squadron we finally came up with a great design. Our idea was to have an RAF grey paint scheme peeling off the aircraft to reveal the markings of an actual Jaguar cat. Our thoughts were that although the RAF was losing its colours there would always be a true Jaguar underneath.

The renderings from Joe were amazing but getting it onto an 11,000kg aircraft was going to be one hell of a task.
We were told that we could use the paint shop hanger at RAF Conningsby but they had a limited window of opportunity, so we had to move fast.

The biggest headache for the squadron having given it the go ahead was the cost. They had a small amount of funds to cover some of the costs but needed additional support. With this in mind we agreed to contact a major magazine to offer them the exclusive rights to the photos. They were delighted and we managed to negotiate the cost of the paint as a result of the deal. Our next target was to contact Corgi and we managed to also secure a deal for the manufacture of a scale model.
With funding in place all that was left to do was book a plane into the workshop!

With approval from our managers the three of us headed off to RAF Conningsby to meet the team and kick off the biggest paint project ever. Alan met us at the gates and walked us around the squadron hangers and then introduced us to Wing Commander John Sullivan. John was the outgoing commander of 6 squadron and would be the lucky one to fly the jet after we had painted it. As the boss I don't think there were many pilots going to argue with him either.

All the team and the jets had moved up from RAF Coltishall in Norfolk for the last months of its time in service. My close friend Mark from Norwich had left the squadron and remained in Norfolk to take up a new career with Bristow helicopters but there were a number of familiar faces that I still recognised, and they all seemed very happy that we were involved. Over the years we had become part of the family and that's what it's all about.

It's always good to be around active jets and I am in my element again. The smell of jet engines and aviation fuel is so very special and you can't get enough of it. Knowing how much I like it the boys take us on an extended tour of the base where I am reunited with the Jaguar T2 aircraft that I few in some years back. I climbed up into the cockpit just to relive that precious moment. Sitting in the back seat brought it all flooding back and I only wish Mike Sears was still around to share it with me.

Having had my personal fix, the guys took us over to see the new kid on the block in the shape of the EJ200, Eurofighter or Typhoon as it was now known. It's a gorgeous aircraft full of modern technology that would turn the old Jaguar inside out. You can't knock the RAF for their actions but many of the pilots believed there was still a role for the Jaguar, and I had to agree with them. It performed exceptionally well in the Gulf war and was amazing at delivering its payload on very low-level sorties.

There were many stories about some very low-level flights but none better than that of Andy Cubin passing by the hangers at between ten to twenty feet at the Omani base of Thumrait. For anybody wanting to watch just how good Andy was then watch Jaguar Part 3 on YouTube.

With the quick tour over we made our way into the paint hanger where XX119 was waiting for us. She looked fantastic sitting in this enormous paint booth and the three of us just stared with raised eyebrows for five minutes.
What on earth had we got ourselves into? Or more to the point what had I got the boys into?

There was no messing about, and we set off to start the prep of the aircraft. With some help from a bit of spare resource we began the task of cleaning and rubbing back the airframe. We each took a section and using your standard wet and dry we started to rub it all back. It was a lot of hard and very dry work but the end of day beer more than helped us to relax.

We arrived bright and early the next day to commence the task of marking out the paint peel affect along the back edge of the wings and then down the front half of the main body. We used up rolls and rolls of masking tape to do this and by lunch time our fingers were numb.

Whilst we were having a break one of the guys asked if we had seen the Battle of Britain memorial flight based at Conningsby. A unanimous 'No' ended up with us all being escorted up through the base to a very special hanger.

With some introductions made one of the senior airmen from the BBMF took us through the corridors and out into an Aladdin's cave of RAF history. At Jaguar we have our own Heritage Centre with lots of our vehicles in it, but this was something else. On either side there were rows of Spitfires and Hurricanes and at the back was the famous Lancaster Bomber. It was jaw dropping stuff for all of us. They ran us through the Spitfires and even let us have a sit in the cockpit which was a real treat. They were absolutely immaculate as you would expect.

We made our way up through the line-up and finished off with tour of the Lancaster. We all climbed aboard at the rear and then climbed through the very narrow fuselage so that we could all sit in the front.

I had been on board the aircraft previously at Coltishall when it had to make a landing and it still had a familiar feel about it. It's not until you have been on board that you fully realise what those brave airmen went through for our country. I sat in the centre gunning tower looking out trying to visualise what it must have been like. The conditions were cramped and with your head popped up out of the top you were a bit of a sitting duck for the bullets of a Messerschmitt. We made our way back through the aircraft and climbed over the main spar before exiting at the rear door. It was a tough job in jeans and T Shirt but imagine what it must have been like with all your flight gear on and large boots trying to buckle up your parachute and get out after you had been hit or were on fire. They were brave men and I salute all of them!

Having had our aircraft fix for the day its back to the paint shop to finish off the tape job. Gary and the spray team stay on late to give the plane its first coat of Jaguar orange so that it can dry overnight.

We arrive back again and we walk into the hanger for a wow moment. Our dull grey Jaguar is now the brightest orange and it looks like it's just been tango'd.

The next big job is to cut the markings of the cat into cardboard so that they can be used as a template.
With about five or six different markings Joe sets off with a can of black paint to do his stuff and I crack on with the tail fin markings and squadron markings. Whilst we do this Gary starts doing the tricky bits of the paint peel effect.

After about four days it's just about done but it's lacking a certain little touch. Like all great masterpieces it needs the artists' signature, but we had been advised that the top brass wanted to keep the jet nice and clean and free from any corporate logos. It didn't take us long to think outside of the box and in less than an hour we had a very special roundel marking that looked like a leaping cat with our initials JDG just below. With nobody around we leapt up the ladders and painted the final marking just below the front edge of the tail fin. It looked right and we just had to do it.

Later in the day the seniors did a fly-by of the hanger and were very impressed with the finished scheme. We had to tell them about our logo and they loved that as well. They were very impressed with how we went around the issue and snook it in.
It was a fitting end to the most wonderful aircraft and a great tribute to all of those who worked on them over the years. We were all immensely proud to have been involved in a little bit of history.

After we had finished, I flew out to America to commence some development testing but I took a call from John Sullivan asking me if I would like to have a flight in a Hercules to see our Spotty

321

Jag in flight. Unfortunately, I would not be back in time but I contacted Joe Buck to stand in and he had an epic experience. They put a harness on him and allowed him to stand on the back of the Herc with the door dropped down while John Sullivan flew Spotty really close so that the RAF photographers could get some photographs. I was gutted to miss out but thrilled for Joe as he did a fantastic job.

After a few outings, to celebrate the time served aircraft, she was finally retired on the 2nd of July 2007.
Along with XX835 and XX725 and in the hands of John Sullivan she made one final flight from Conningsby to Cosford where she remains today as a running aircraft to train future technicians and ground crew.

It was the first time the RAF had commissioned a paint scheme for an aircraft and it went down well. I didn't realise just how well until I went into W.H Smiths to collect my copy of Aircraft Illustrated. I grabbed two copies and went to pay at the till. The young lady admired the front cover with a pic of Spotty and commented that her fiancé loved the plane. I mentioned that I had been involved in the painting of it and then showed her the pics in the book and on my phone. She was amazed and told me to hang on a moment. When she returned, she had a couple of copies and asked me if I could sign them for her fiancé. It was so cool although my wife was just embarrassed by it all.

I returned to Cosford on a couple of occasions to support some air shows and managed to get some great pics with the old girl and one of our F-Types. She has also become a celebrity at Cosford and is often out on the runway for plane enthusiasts to take pictures of.
Corgi finally released a model of Spotty and a couple of plastic plane kits have also been released. Obviously I had to buy them all.

Sepecat Jaguar's first flight 12th Oct 1969
Spotty Jaguars final flight 2nd July 2007
Enjoy your retirement old girl.

# 28. The Red Arrows Also Love Cars

I have admired, loved and watched the Red Arrows for as long as I can remember. They formed in 1964, the year I was born, and started flying the Gnat through until 1974. As a young boy I remember seeing them on TV and at a local air show. I also recall taking a trip out to RAF Bitteswell near Lutterworth when they were being serviced.

In 1974 they changed over to their iconic Hawk Aircraft, which they still fly today.

I have lost count of the number of times I have seen them fly but the most memorable displays were when they joined up with Concorde. The flights in 1996 for the 50th anniversary and then 2002 for the Queens Jubilee were quite stunning.

I never dreamed that I would end up getting very close and personal with the team, but a chance request from an ex Jaguar pilot would change all of that.

When I flew in a Jaguar jet my wingman for the day was Flight Lleutenant Graham Duff (Duffy). Duffy flew Jaguars for many years before moving on to become an instructor at RAF Valley flying Hawks.

Graham had contacted me to see if I could set up a day visit for his students at Jaguar Cars as part of a documentary or something like that. They came over and I took them around Whitley and Gaydon and then did some flying laps in a Jag on our test track. It was quite interesting to see these fighter pilots jumping about in the cockpit of my XK8 and then screaming as we threw it around the circuit. I guess being fifty feet from the floor in a 600mph jet is quite scary but even for them being a few inches off the floor at 100mph is even scarier when they are not in control.

We had a great time and as result of my efforts Duffy arranged for me to join him for a flight in his Hawk from RAF Valley.

Having experienced my dream of flying a Military Jet with 6 Squadron I never imagined I would get a second go. I went over to Valley and was kitted out for a training run around North Wales that went into some of the Mach Loop as well as going ballistic for some aerobatics. Even though it was very bumpy we had an amazing low-level flight through the valleys and then we went vertical through the clouds into the clear blue skies. As we went over the top, I felt my first bit of weightlessness and even let go of my camera for a couple of seconds so it could float in the cockpit.

Duffy was an amazing pilot and instructor, so it came as no surprise when he joined the Red Arrows in 2008 for a three year tour of duty.

It was whilst he was at Scampton that we made contact and I was asked to support a STEM event with the team.

STEM (Science Technology Engineering Mathematics) is supported by JLR and many other companies to help educate school children. Both my wife and I had completed a couple of STEM projects through work but being asked along to a Red Arrows event was a big one!

I took a drive up to Scampton to meet the team and get an idea of what we were going to do as well as taking up the invitation for a tour around the base.

Scampton was built in 1916 and used by the First World War Royal Flying Corps until 1919. In 1936 it was re-opened for bombers to use in the Second World War and then Vulcans during the Cold War. It was temporarily closed in 1996 but was reactivated to house the Red Arrows.

Its biggest claim to fame came on the night of the 16th to 17th May 1943 when 617 Squadron flew out for one of the most famous missions of the war. A total of nineteen Lancasters left Scampton under the leadership of Wing Commander Guy Gibson and the team became known as 'The Dam busters'

Having had a fantastic tour around the base and confirmed what we needed to do for the event I went to leave but I couldn't get anywhere near my car as it was surrounded by aircrew. I had taken up our development XKRS and they loved it. I spent another thirty minutes answering questions about the car and giving them all the performance data and then found some additional time for one or two passenger rides. Although they get to fly jet aircraft every day they still love cars!

At the actual event I set up a static display of a couple of cars as well as getting some remote-control Land Rovers that you had to negotiate around a specially designed circuit. We had about three hundred local school children attend, and my role was to take them in groups to talk about Engineering using the cars as props. After my talk they then got the chance to show off their driving skills with the remote-control Land Rovers.

The event was also supported by Mercedes Formula 1, Aston Martin, Vulcan to the sky, BAE Systems as well as many more Engineering Companies. We were also joined by Professor Brian Cox for the Science discussion and the kids loved his presentation. Unfortunately, we were supposed to be joined by Carol Vorderman for the Maths, but she couldn't make it. In subsequent years we did get Rachel Riley from Countdown for the maths bit. She blew us away with her knowledge but her two best attributes were actually her love of Jaguar Cars and Manchester United. I did three STFM events for the Red Arrows taking up the Austin Powers XK8, my XKRS-GT, Project 7 and an F-Type SVR and whilst the kids loved seeing them it was the Air Crew that got the biggest buzz from having them there.
They were all enthusiastic petrol heads and any opportunity to have a go was accepted. The runway was normally closed for the rest of the day, so it always seemed right to share the fun and let the squadron experience some proper acceleration off the line.

It wasn't just the squadron who enjoyed the experience either. Over the three years I had some excellent assistance from my closest work colleagues in our department. Pete Page, Mark Twomey and Paul Raisebeck were all willing volunteers to talk about our great cars and I'm sure they would all agree that the events were a major success and we all hoped that some of the young guests would be inspired to take up engineering in some form from what they had seen and been involved with.

It was during the later events that I became a good friend of Flight Lieutenant Marcus Ramsden. Marcus was in charge of the support crews that you see at the air shows. They all wear blue overalls and each Red Arrow is assigned his own technician. They are referred to as 'The Circus'. Each engineer is assigned to a Jet number and Marcus was Circus 1 looking after Red 1 (The Boss).

Marcus was passionate about sports cars and even did a bit of kart racing so he always looked after the Jag lads.

During his time as Circus 1 he would always send me updates of where the team were and what they were doing. He even went the extra mile to send us pictures from the back seat and video clips as he knew how much we loved the aircraft. It was a relationship that for some reason always works well. Pilots and crew love cars and we love the jets.

Having supported their events over the years I was invited up to the base for what turned out to be the most amazing days you could ever wish for. Marcus invited me and my wife up for a private tour but when I said I had a very enthusiastic six-year-old daughter who was mad on the Red Arrows he invited us all along.

We turned up at the security hut to book in and to our surprise Olivia was booked in as a VIP guest. She had her picture taken in her own Red Arrow suit and was given her own pass for the day. She felt very special.

Marcus met us and took us up to the hanger to meet his team. Olivia was taken around the jets to meet the Circus Team and they treated her like royalty. We then went into the flight ops to see all the maps and the planning side of the business before being taken for tea and biscuits in the crew room.

It was a working day for the Reds, and they were about to do a practice run of their new routines so we were thrilled when Marcus said we had been invited to go in and listen to the flight brief.

As the Reds entered the crew room everyone stood up and they made a massive fuss about Olivia being a VIP guest. Because she was wearing her Red Arrow suit, they said she could sit in the red chairs at the front with the team. Mother and father had to sit at the back on the blue chairs that are reserved for the technicians.

It's not a massive crew room so being at the back was no big deal and we were still close enough to see the biggest smile on Olivia's face.

Olivia sat with Red Ten who was Squadron Leader Mike Ling. Mike had an amazing career with the Reds having had a record-breaking time with the Squadron. He became the longest serving pilot in the team with ten years' service.

As Red Ten he became the teams' supervisor and also provided commentary at all the air shows. His flying stats are quite impressive with 2585 sorties, flown 1765 hours, visited 46 countries and commentated on 699 shows.

She was certainly in good hands and when they left the crew room she proudly walked along with the team to join the jets outside.

The team got themselves sorted out and Mike took Olivia out onto the grass where they had a video camera set up to film the practice. He also had a radio so that he could talk direct to the pilots and even let Olivia say a few words before they departed.

I'm not quite sure how Olivia was feeling watching them all take off, but I was in cloud cuckoo land. They took off in groups of three and the noise was fantastic. Just seeing them take off makes you smile, and you do get an overwhelming sense of pride.

With the jets disappearing from the airfield Mike asked Olivia if she would like to pose for a photo. He moved her around a bit until he was certain he had the best spot and then moved away just in time for a synchronised low-level pass over her head. It was just the coolest thing you could ever wish for. If they had done this for me, I would have been delighted but because they did it for my daughter it was extra special. She will remember that moment for the rest of her life!!

We watched the whole practice and then greeted the entire team back when they parked up all the jets outside the hanger.

The pilots made their way back to the crew room to watch the video of the display to assess their performance. We were again invited back in to watch the review. Unfortunately, the day was all too much for the six year old Red 11 who collapsed in my arms much to the amusement of the team.

It was an eye opening debrief with the most open and honest review of individual performances I have ever seen. The attention to detail was on a different level to what I had imagined and you can see why they are the best of the best.

It was a fantastic day and I know just how much it inspired Olivia. She went back to school and stood up in class to tell them all about her day, and even today she will still tell people that she wants to become a Red Arrow pilot.

The entire squadron were just amazing and a credit to both the RAF and the country that they serve.

My special thanks go out to Flight Lieutenant Marcus Ramsden for the invite, Squadron Leader David Montenegro for allowing us to be part of the team and to the legend that is Mike Ling MBE for being so kind to Olivia and looking after her.

And how did we end up doing this....... because The Red Arrows also love cars.

# Part 6 – So what's It All About?

Having written twenty eight chapters about a lot of cool stuff I just had to ask myself the question. What's it all about?

Quite clearly as a business it's about selling cars and making a profit but it's so much more than that. It's about the brand, its heritage and passionate people who have made it that way.
It's also about the cars and the impact they have on us and the customers.

In Chapter twenty nine I wrote about the time I had to replace an older prototype with the new model and how it made me feel. I wrote this piece for the XF Vehicle Office Team because it was our favourite car.

We spend a lot of time in our cars just going from A to B but when you get into a special car how does it really make you feel. I have been lucky to drive a lot of them and in chapter 30 you will see that one of them made me feel very special!

In chapter thirty one I try to finally answer the question, 'What it's all about.'
Living the dream and leaving a legacy.

# 29. A Vehicle Love Affair

I wrote this for the XF Team when I exchanged our old development car for the new version and I think it captures just how I felt at the time. It's a vehicle love affair thing I guess!


It was indeed a sad day today for engineering love affairs.

As passionate engineers we often get attached to fabulous looking cars that make your hearts race. It's a bit like that first kiss behind the bike shed as a spotty teenager but unfortunately you can't go back to relive that moment...... You just have to move on.

The girls certainly get older and more exciting, and as you build up more experience you do tend to go for more thrills, but you often catch yourself thinking back to that first moment........a bit like I did today when I had to take VX57XFR back to Whitley from Gaydon.

Together we had many thrills during the first days of development right through to the glamour of the launch, but today was different.

As I walked down to the porta cabin on a wet and windy Friday there she was, waiting for her old flame. I slipped inside full of anticipation that the sparks would fly, as they did on that Glorious Day in Devon when we filmed for Top Gear.

As it was Friday I thought she might like to go the back way, through Cubbington and Stoneleigh to dance a sizzling salsa on the wet tarmac.

I caressed the steering wheel and fumbled for that magic button that would unleash the fire in her heart. The old girls motor was definitely willing, so we set off gliding over the tarmac covered ballroom.

As we embraced the moment I started to notice a few things that bothered me. There was some cosmetic surgery, a bit of wear and tear and the early signs that the old girl had been round the block a bit. The stark realization that she had been abused over the last few years made my heart sink, because suddenly we're not doing a Salsa anymore.......we had slipped gracefully into a ballroom waltz.

She was trying her hardest to please me, but it just wasn't the same. She was pleading with me that all who had driven her over the last couple of years meant nothing to her, but alas....it did to me.
In her day she was Purdy, the New Avenger....... but now she is just Joanna Lumley...still elegant and a little racy, but finding it difficult to recapture her youth.

Arriving at Whitley I remembered some glorious moments......the What Car Awards, where she looked amazing, the front cover of EVO and Autocar and her starring role in Top Gear, and after all those moments she was always willing to party on through the night......but not today.  Although her old flame was ready to go to third base all she wanted to do was rest in a cozy spot behind the PME building.........not loved like she used to be but happy.

But the night is still young, and in the parking bay by the design centre is the new girl on the block.......550hp, 680Nm of torque and a whole lot of attitude......the new Purdy........a Michelle Rodriguez of the car world.

We depart ready to conquer the world, but I can't help but look back as she sits there, tears of rain dripping down her headlamps......she was probably thinking , it's better to have loved and lost than to have never been loved at all.

# 30. Cars Make You Feel Special

Every day you get in your car and go backwards and forwards to work in a groundhog sort of way. Nothing special just a daily commute from A to B and back again. But every now and again your 'Groundhog Day' is interrupted with the chance to drive something different, something special: an F-Type and the effect it has on you is quite awesome.

As I stroll out and look over at the car, I have the same schoolboy emotions that I used to get when walking into the paper shop to look up at the adult mags on the top shelf. You start to stare in a dream like haze as the blood moves down and away from the brain and takes up a permanent residence in your pants. Its pure automotive porn.

As a schoolboy I had to wait many years before I could enjoy the delights of the top shelf but today it was just like my dad had walked into the shop, bought the magazine and then passed it down and said "son, it's all yours. Fill your boots and enjoy the ride." So, with sweaty hands, a box of tissues and a book of dreams I make a run for it before anybody changes their mind.

The F-Type is now up and running and my childish brain is already pleading with my right foot to press the pedal and make a lot of noise. You just can't help it.

Throughout my younger years I have been obsessed with noise. Rock music from ACDC, loud motorbikes, jet aircraft on full re-heat, big expensive fireworks and thunderstorms. Today I have to add the F-Type's exhaust soundtrack to the list. It's impossible to drive away and let it auto shift. You have to engage sport mode, paddle shift and give it the whole nine yards of delicious overrun. The result is something akin to pouring a lorry full of Rice Krispies into a vat of aviation fuel: SNAP, CRACKLE and POP POP POP.

The car has already rekindled the feelings of my younger years but as you motor on it transforms you into a different character. I'm now flicking the paddle shifts, lifting off, downshifting, accelerating hard planning my route down the fosse like a formula 1 driver and I'm loving it. The car makes you feel like a world champion and even though you have an exhilarating exhaust noise you still end up making F1 type noises as you blast past your competitors (sorry, fellow road users).

As I approach the next town the F-Type transforms me again. Paddle shifts off, automatic mode engaged, sunglasses on there you have it: a seamless change from F1 driver to celebrity football player. I take a quick glance into the shop windows to confirm that yes, I am David Beckham. The noise of ACDC has been replaced with The Red Hot Chili Peppers and the high street crowds are standing up to applaud my appearance.

How can you not enjoy this? I crave more so the route goes into extra time with and extended drive to get some food. What would David Beckham do? Aldi? Nope. Morrisons? Nope. Waitrose? I think so.

In doing a drive by of the car park there are no suitable spaces available that are worthy of a superstar. I end up doing three laps until a primary slot is available.

Briefly I just have to return to schoolboy mode and give the throttle a nice blip to ensure all the heads at the checkouts turn to look. Walking away from the car there is an approving nod from a fellow player, he glances back at the F-type and in my mind, I'm already shouting: "Yes!"

With some celebrity meals, courtesy of Charlie Bigham, in my bag I set off again.


I crackle out of the town centre and back into the countryside with the top down and with a warm breeze brushing my hair there is yet another seamless change from Beckham to Bond and before you notice it The Red Hot Chili Peppers have been replaced by Adele.

Suddenly in my rear-view mirror I see a black BMW. It's got to be the German villains. Here we go, buckle up Adele I'm just about to make the Sky-fall down on you. In a couple of seconds waftmatic mode is replaced with Secret Service Sport Dynamic Mode and the quad pipes are spitting firestones at the chasing BMW.

With the Germans swiftly dispatched all too easily Bond disappears and returns to the tranquility of the Leicestershire countryside and a little further up the road there is a horse and rider so the gatling gun exhaust mode is switched off and normal mode engaged to promote a classy and elegant arrival into the village.

Slowing down to walking pace horsey lady looks down at Bond and his trusty steed. No words are required at this point, just mutual respect for everything that is horsepower.

The drive is almost at an end but there is just enough time for another mode change: Daniel Craig has left the vehicle to be replaced by George Clooney. I am a movie star of stature, fame and wealth. On the outskirts of the village I pass the 4x4's and park up outside the mansion. The neighbor's exchange knowing looks: "this man is class," "this man fits in here." I nod in agreement. "Yes, I have arrived. I am the main man and I appreciate your attention."

In a thirty five mile trip you don't need to worry about all the technical stuff. Who cares about on centre feel, do you ever worry about secondary ride inputs or throttle pedal feel? Did horsey lady question the transition to oversteer in the wet? I think not.

What did the F-Type do to me? I will tell you

It transformed me into whoever I wanted to be and made me feel alive.

Before I went to bed I looked out of the window and glanced down at it. The summer moonlight cast gothic like shadows over the black paint and as it did, I thought to myself: "Oh yes. Tomorrow morning, I am Batman."

# 31. Living the Dream and Leaving a Legacy

It's taken me a number of years to get to this point in my life and my book. I have been thinking about writing it since my epic drive through Australia in 1998 and then when I read a Bill Bryson book about his travels in Australia and figured that my own journey was far bigger and worthy of a read. Unfortunately, life and work takes over and I never found the time to do it.

After a lot of my subsequent travels and tests I kept thinking about additional chapters and I even drafted out a few ideas over the course of a quiet Christmas, but it wasn't until I had children later in life that I suddenly realised why I needed to write the book. My children had missed most of my career and they would only know stuff about their dad from the later stages.

Most people of my age have had their families in their twenties or thirties and the children grow up through the significant periods of their parents' lives. We had Olivia when I was 47, and at that age I was already thinking about my pension. Jessica joined us when I was fifty and then Orla completed the hat trick when I was fifty five. Luckily their mother is a bit younger which helped me in the decision to have children in the first place. It was a major decision that was made easier by the adage, 'You're only as old as the woman you feel' and that is so true! Apart from the time that.........well, that's another story!

With the addition of children in my life it made me think more about what I had done and achieved that they had missed out on sharing. As a youngster I sat listening to stories from my grandparents about their lives, but they always got mixed up with the details and timings, so you were never sure what was fact or fiction. I didn't want my kids to go through the same sort of stories when they got older with their poor old dad starting to lose his marbles. I'm sure they would sit and listen to all my adventures just to make me happy, but how much they took on board would be open for debate. Getting it all down into a book however, whilst I'm reasonably sane and have a good grip on my marbles, will hopefully give them something to cherish and pass on to their own children when I have long gone.

So, what sort of legacy am I leaving them in my written words?

Well, I haven't had an impact on World events, I haven't invented anything of significance and I certainly haven't been worthy enough of making it into Wikipedia or a good Google search! So, what have I done?

When you walk into the Whitley Engineering centre there is a bronze statue of Sir William Lyons who was the founder of Jaguar cars. He left a massive legacy but there will be no statue for Dave Moore, unless I come up with some alternative fuel system that makes JLR the richest company in the world in the next few years. You don't have to dig deep in Jaguar's history either to find employees who have all left a legacy. Malcom Sayer designed the most gorgeous cars on the planet in the C and D-Types that both won at Le Mans, and in the 1960's he designed the gorgeous E-type. He was joined by Bob Knight who designed the chassis, William Heynes, Walter Hassan and Claude Bailey who were the engine designers and Lofty England who ran the motorsport division.

There were the drivers too, Duncan Hamilton, Tony Rolt and Sir Stirling Moss whose names have been immortalized in the history books. Moving away from the 1950's there are more recent employees who have left their own legacy. Designers such as Geoff Lawson and Ian Callum have left us with some iconic vehicles that will be retained in our heritage museums for many years to come and Chief Execs like Sir John Egan and Nick Scheele have guided us through good and bad times to get us to where we are today.

So, going back to the original question, what have I done?

Well, if you have read and enjoyed the previous thirty chapters you can see that I have had a blast doing a lot of stuff. In fact, a lot of amazing stuff. I don't really know how I got to do it all but I must have been doing something right?

I left school with just the basic qualifications. I tried my best and worked as hard as everybody else, but it just wasn't my time. I played every sport for the school, made friends with everybody and never had any enemies. I ended up with an amazing school report and in the end I decided to go into engineering.
As I matured the timing was right and I finally achieved my rewards at college and found success at work.

With experience and confidence, I moved to Jaguar and enhanced my position with a degree as a mature student. I found it easy because I already had a degree in life and what I was learning became more relevant. I'm not sure if this was the right way to do things or not but it worked.

I am currently reading a book by Michelle Obama and she has been quoted as saying "There is no magic to achievement. It's really about hard work, choices and persistence", and I guess that's true for what I have done. I had to work hard before I got my achievements, I had a choice and decided to be an engineer and with persistence I joined a company I was passionate about.

So, is it passion that gets you involved with all the stuff? It certainly helps that's for sure because I have not met anybody who has not admired it, but there are many people like me with bag loads of passion that don't get to do anything like this or turn their dreams into reality. So, what is it?

Well, it's all about making friends and being part of a team. It's about creating energy and enthusiasm that your friends and teams can draw on. You don't have to be the centre of attention and it's not about what you have accomplished in your career or life. It's more about what you inspire others to do that really matters and if you really love what you are doing you will be successful which has to be the very essence of my legacy.

In all that I have written I have been living the dream and that is all I want my girls to do. To have passion in what they do and take the right path to turn their dreams into success.

So, to my gorgeous girls, Olivia, Jessica & Orla.

I have had my journey and now it's time for yours. Dream about what you want to do, have the vision to find it, believe in yourself and just do it.
Enjoy your own journey with a passion and have a blast along the way as well.
I'm sure you too will all have an 'Epic Journey Fuelled with Fascinating Events'

Love Forever & Always

Dad x

Printed in Great Britain
by Amazon